Henry Hall Dixon

Field and Fern or Scottish Flocks and Herds

Henry Hall Dixon

Field and Fern or Scottish Flocks and Herds

ISBN/EAN: 9783337106867

Printed in Europe, USA, Canada, Australia, Japan

Cover: Foto ©berggeist007 / pixelio.de

More available books at **www.hansebooks.com**

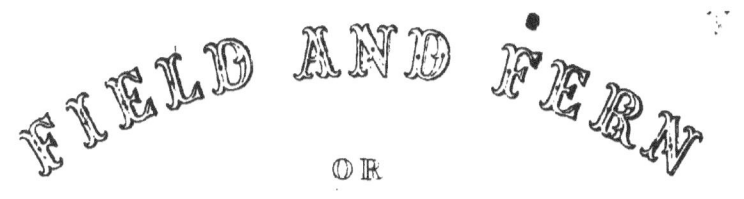

FIELD AND FERN

OR

SCOTTISH FLOCKS & HERDS.

(NORTH)

BY H. H. DIXON,

AUTHOR OF THE R.A.S. OF ENGLAND PRIZE ESSAY ON SHORTHORNS, 1865.

DEDICATION.

TO

ARCHER FORTESCUE,

IN REMEMBRANCE

OF SOME VERY HAPPY DAYS, UNDER HIS ROOF

IN THE ORKNEY ISLES,

THIS BOOK IS DEDICATED BY HIS FRIEND,

THE AUTHOR.

PREFACE.

"Then hey for boot and horse, lad,
And round the world away:
Young blood will have its swing, lad,
And every dog his day."

PROFESSOR KINGSLEY.

"HE came back with a blood filly which had put out a curb, half-a-crown in his pocket, and his hat stove in." Such was the fraternal narrative of the return of a prodigal, who had gone leather-plating for a season. Read in lieu of the above an Orkney garron, just four-pence out of a good round sum, and two fat, little note-books filled to the gorge, and it pretty nearly describes my belongings, when I reached home in the snow of a February night.

The original idea had been quite royal—"My pleasure in the Scottish woods three summer months to take." After working hard among the flocks and herds of England for four years, I was naturally anxious to be over the Border, and find new scope for pastorals. I wished to visit past and present Highland Society winners, in their own stall or fold, and to gather evidence from those breeders who stand high in its annals, not only as to the present progress of the stock on which England depends for such extensive supplies, but also as to the thoughts and labours of men who have done Scotland good service, and then passed to their rest. Grouse shooting, deer stalking, and salmon rod-fishing have their own liege lords of the pen; but still there were many little points connected with hunting, coursing, racing, and otter hunting, which seemed calculated to work into a picture of Scottish life, and to vary the monotony of mere beef and mutton chapters.

Fancy soon faded into reality, and I found that I had set myself a very serious task. I had to pluck the heart out of three summers, a winter, and a spring, to travel some 8,000 miles,* to sleep away from home about 250 nights, and change my bed 146 times, before I wrote a line. The Government Fish Commissioners, coasting jauntily along in the *Salamis*, had quite the best of me, as I worked my own commission on Flesh and Fowl, through sunshine and shower, with no secretary to cut out the line. It is very easy to draw up a programme, but not so easy to hold to it. I often found a new and valuable witness where I least expected to do so, and had to throw over every plan rather than leave him; but there was still, in spite of all the hardship and

* This includes several journeys to and from London.

harass, quite a pleasant soldier-of-fortune feeling in never being sure whether you would turn up at night by the fireside of "a golden farmer," or in a hole in the wall at a wayside inn. Mere scenery I was obliged to disregard. In fact, it was of no use to me, unless it served as "setting" for some crack sheep or cattle; and acting on this purely-practical view of things, I sternly held my line, regardless of the most glorious combinations of water, wood, and mountain, for which other tourists were ever turning aside. I did not even spare a day for the Trossachs, but went "hot trod" past the guide post after black faces towards Rob Roy's grave; and my eye might never have rested on Killiecrankie, if I had not passed through it on my way to the West Highland herd at Blair Athole.

"Something attempted, something done,
Has earned a night's repose,"

was my motto, and I enjoyed one between two and four a.m., in the saddle, during a night ride over the Ord of Caithness, while the rain poured and the mare grazed. "Cockade"—so called from persistently wearing her mane on the near side—was not my companion in the summer of '62. I thought at first that I would walk, but it was a great mistake. It may answer for a mere light-hearted saunterer, who wants to take a few sketches, and his ease in his inn, but not for one who has a responsible task in hand.

Coaches and railways aided me in a measure, but I wearied sadly under a very heavy knapsack; and such long cross-country walks were not especially favourable to framing cross-examinations at night. Hence I soon found that I was merely cutting time to waste, and, after making the discovery, I pushed my way to the Orkneys, just to get a notion of the work before me, and asked my good friend Archer Fortescue to buy me a garron before that day twelve months.

Another summer came round, and there were only two garrons of the size for sale in Pomona—one at £10 and the other at £7 10s. The brown was just the thing, although it was rather ugly; but the bay looked, when I met it by chance on the deck of the *Vanguard*, as if it would have come in half with me. Condition was everything at such a crisis; and, thanks to "Moore's Dietary of Corpulence" (which is very nearly the same, but several years senior to the "Banting system"), I was enabled to take 24lbs. of flesh off my back, and carry it behind me in the much pleasanter shape of macintosh and luggage. *"Just fifteen four the lot,"* was the announcement of Provost Bell of Dumfries, when, grasping my pad, valise, book-bag, and macintosh, I sat in his bacon-scales. Many and various were the suggestions about saddles; but a pad seemed best, on three grounds: it would fit almost anything if my mare died or was disabled; it was far more easily carried; and as it folded round the valise, it sometimes served for a pillow on the heather.

"*He'll never get to Lunnun, maister,*" said Dick, the first whip and kennel huntsman to the Orkney Harriers, *sotto voce,* as I took the mare from his hand in the Orkneys; and I was not quite sure on the point myself. Because we didn't go with him from Kirkwall to Wick, Captain Parrot will have it to this hour that we swam the Pentland Frith, just by way of a relish at starting. The journey, to a man who has a good horse and can send his luggage on to points, must be a remarkably easy and pleasant one; but when you have only a shy half-bred nag quite out of condition, and have, perforce, to spend so many months roughing it, in a country to which you are not acclimatized, it becomes no May game. Still, with fine weather, and a steady practice of getting off to lead for every third or fourth mile, it is a grand independent way of travelling. I may say it was positively exhilarating to put the mare's head straight across Scotland, during a hard frost, from St. Boswell's to Ayr, and cut down the hundred miles at four-and-forty a day; or to rattle from Athelstaneford nearly to Kelso over the Lammermoors, with two shirts and three pair of stockings on, and the cold cutting your cheeks to the bone.

Being asked "*How's your wardrobe?*" &c., as you

ride through a town, is as nothing; but there were sundry disadvantages connected with this ancient mode of locomotion. It is a weary thing sitting three-quarters of an hour on a corn-box at night, to be sure that the ostler does you justice. Every ferry-boat in the Highlands was fraught with a fresh difficulty, and even the master-minds of Meikle Ferry quite thought that they must have sent me many miles round by Bonar Bridge. Every railway-train produced a fresh run-off; and I was lucky if I could put my mare's head in the right direction, so as to get a three-hundred-yard gallop to the good. It was equally objectionable having to blindfold her and stuff her ears, and twist her five or six times round, to make her forget which way you wanted to go, when you found a Lanarkshire or Ayrshire blast furnace roaring like a lion in the path, late at night, between yourself and your inn.

Still, all these were very minor troubles in comparison with the collection and sifting of book materials. Most Highland places seemed to be spelt in two if not three different ways; and the Gaelic names of bulls and cows almost drove me to despair, even with the Gaelic dictionary at my elbow. After all my labours, the most that I can lay claim to is to have given a general sketch of Scottish farming from that prize-stock point of view which is being gradually worked out so ably in all its details, not only by those which make agriculture their speciality, but by local newspapers as well. I have already profited not a little by their labours when I compared their notes with my own; and I have drawn many a hint from the Transactions and Records of the Highland Society, whose Secretary, Mr. Hall Maxwell, has lent me, both in this and other respects, most invaluable aid.

To ensure accuracy as far as possible (though I see with regret that I have not given the Marquis of Tweedale credit for the first private introduction of steam-ploughing to Scotland), I have not sent a sheet to press without previously submitting it to those most conversant with the herd or the district, on precisely the same system that all witnesses before a Parliamentary Committee receive their evidence to revise. As regards the vein of sporting, which runs more especially through the "South" part, I may mention that the whole of the coursing was kindly looked over for me by that eminent ex-judge, Mr. Nightingale, and that the quoted descriptions of the styles of many of the great winners are nearly all from his lips. To him and scores of other friends, who have cheered me on in my labours, and greatly smoothed my way by their hints and hospitality, I owe a very deep debt of gratitude.

I originally named, and in fact advertised this work as "Field and Fold," and then found that the

Religious Tract Society had already issued a sixpenny publication of that name. Perhaps, however, "Field and Fern" has a more strictly Scottish application. The division of it into two independent parts, "North" and "South" of the Frith of Forth, seemed most natural, and calculated to meet the wishes of such Highland and Lowland purchasers as might have no interest in each other's stock lore.

As readers never by any chance look at a table of *errata*, I have adopted a totally new plan, viz., correcting any little thing that specially called for it in a foot-note to the text, when I saw an opening in the course of the work. Six or seven notes of the kind will be found. I may also add that I have used the name of the parish "Coultar" when I ought to have said "Culterallers," that "thin *for* plantations" in reference to the Renfrewshire country should be "thin *fir* plantations," and that "Edinburgh town" has crept in for "Edinburgh toun."

As regards the portraits, I have chosen Mr. Hugh Watson, Professor Dick, Mr. Nightingale, and the late Duke of Richmond as representatives of the cattle, horse, greyhound, and sheep interests. Mr. Gourlay Steell, R.S.A., has kindly presented me with the head of "Duntroon," one of those Highland chieftains of the heather, which will long survive their sirloins on his canvas. The Master of the

Teviotdale sits among his equally hairy darlings, with his Lord Chancellor "Sandy" at his side; the scene at Knockhill typifies the Turf, the Leash, and the Chase in Scotland; and my own mare stands hooked to an out building, and, to all appearance, quite resigned to her sadly vagrant life.

From first to last, this work has been very nearly three years in hand; but, spent as much of them has been among such new and varied scenes, they seem to comprise a lifetime. No one but those who have been regularly "in the mill" can tell how difficult it is to reconcile and winnow conflicting opinions given by men of mark on the same point, and to put some light and shade into the history of flocks and herds, which has an infallible tendency to degenerate into mere vain-glorious invoice-lists of males sold and prizes won.

"Men have no faith in high-spun sentiment,
Who put their trust in wedders and in beeves;"

and no one would "try it on" with them. Still, on the other hand, it is only just that readers should remember that an author who is obliged to put such very matter-of-fact objects as "wedders" and "beeves" in his foreground, instead of human beings, with their joys, and their sorrows, writes at fearful odds, and has virtually no scope either for language or fancy. Hence, in racing phrase, he is clearly "entitled to claim an allowance."

However, the book is done, after many interruptions from illness and other causes; and I seemed to breathe quite freely when I signed the last proof-sheet. I can only trust that it may prove to me the little scarlet pioneer of a still more extended tour through England, Ireland, and a portion of the continent; but go where I may, every August will bring with it the old yearning to be across the Tweed, and all the pleasant memories of my journey

"From the Orkneys to Kensington with Punctuality and Despatch."

H. H. DIXON.

10, *Kensington-square, W.,*
August 1, 1865.

CONTENTS.

CHAPTER I.

ABERDEEN TO THE SHETLANDS.

Departure from Aberdeen—Voyage—The Shetlands—A Lively Passenger—Cabin Councils—Arrival at Lerwick—A Peep at the Town—Walk to Voe—The Lonely Muirs—The Cows—The Ponies—Pony Life in the Coal Mines—The Sheep—Wool Manufactures—Improved Stock in the Shetlands—Back to Kirkwall . . . 1—20

CHAPTER II.

THE ORKNEYS TO THURSO.

Orkney Nicknames—Shapinsey—Defence of Hellersay—Orkney Panorama—Mr. Balfour's Shorthorns and Shetland Ponies—The Legend of "Spunky"—Orkney Sheep Crosses—Sheep Marks—Devon and West Highland Bulls—"Fishing Pork"—Orkney Garrons—The Fortescue Harriers—Swanbinster—Postal Difficulties—A Sea Sick Horse—The Message from the Ice—The Hoy Farmer Consoled 21—39

CHAPTER III.
THURSO TO HELMSDALE.

The late Sir John Sinclair—Caithness Sheep Farming—St. Mary's Mass—Georgemas Tryst—Shorthorn Crosses—Barrogill Castle—Barrock Plantations—Shorthorn and Galloway Crosses—Bringing-up of Calves and Yearlings—Sir George Dunbar's Leicester Flock—Mail Journey along the Coast—A Night Ride on Horseback . 40—53

CHAPTER IV.
HELMSDALE TO MEIKLE FERRY.

A Walk down Kildonan—The Cheviot Flocks of Old—Foxes on the Gibbet—Mr. Hadwin's Stags' Heads—Gideon Rutherfurd—Kildonan Churchyard—Sutherlandshire Flockmasters—Shows in the County—Sheep Farming—Heather Burning—Sporting in Sutherland—Strath Brora—The Dunrobin Herd—The Museum and Kennels—The Dunrobin Dairy—Climbing Ben Vraggie—The Meikle Ferry , 54—72

CHAPTER V.
TAIN TO INVERNESS.

The Black Dog of Tain—A Jockey Club Wanderer on Morich Mhor—Ross-shire Tod-hunters—The Crofters—Flocks in Easter Ross—From Tain to Dingwall—Coul Cottage—The Black Isle—Belmaduthy—The Muir of Ord—The Caledonian Boots—Highland Society's Shows—Inverness Character Fair—Hugh Snowie's—Stags' Heads—Laggan Cottage—Nairnshire—Lord Cawdor's Old Scotch Flock—Kildrummie and Lochdhu—The late Hillhead Herd—The Witches' Heath
73—94

CONTENTS.

CHAPTER VI.
FORRES TO FOCHABERS.

The haunts of St. John—Morayshire Sellers—"Horned Beasts"—Lack of Breeders—Feeding Stock—Mutton for London—Sheep Farming—Morayshire Feeders—The Forres Fat Show—Altyre Stock—The Balnaferry Herd—From Forres to Elgin—Pluskardene Abbey Bargains—Wester Alves and Ardgay—The Story of the Buchan Hero—Elgin—The Spey Fisheries—The late Duke of Richmond—Gordon Castle—Its Breed of Setters—Its Flocks and Shorthorn Herds . . . . . 95—113

CHAPTER VII.
FOCHABERS TO SITTYTON.

A ride from Fochabers to Aberdeen—Back to Orbliston—A Shooting box—The Mulben Herd—The Pig Trade of Banffshire—The Portsoy Cartsire Stud—Clydesdales—Banffshire Shorthorn Beginnings—The Rettie Herd—Mr. Rannie's Leicester Flock—The Montbletton Herd—The late Mr. Grant Duff—His Catalogue Notes—The Forglen Breed of Cows—Mr. Lumsden's Herefords—Hereford Crosses—The Kinnellar Herd—The Sittyton Herd—Udny and Jamie Fleeman
114—145

CHAPTER VIII.
SITTYTON TO ABERDEEN.

The Aberdeen Fat Cup—Mr. Stewart's Cracks—The Tarty Herd—Mr. Martin's Show Beasts—Aberdeenshire Feeders—Early Days of Feeding—The Original Horned Doddies—Aberdeen Meat Supplies—The Cattle and Dead Meat Trade with the South—The M'Combie Family—Easter Skene—A Day at Tillyfour . . 146—181

CHAPTER IX.

ABERDEEN TO STONEHAVEN.

The Royal departure from Balmoral—Up the Deeside—Kincardineshire Sheep-feeders—The Portlethen Herd—The last of Fox Maule—Colour Conception—The late Mr. Boswell of Kingcausie—His Highland Society Essay—From Bourtree Bush to Stonehaven—The late Captain Barclay—The Old Days of the Defiance   182—210

CHAPTER X.

STONEHAVEN TO CORTACHY.

The Weathercocks of Stonehaven—The Sea Coast Road and The Mearns—Stock of the District—Bulls at Fernyflatt—Yorkshire Calf Trade—Angus Commentaries—The Kinnaird Valley—Old Montrose—Black Herds—The Kinnaird Castle Steading—Druid and Cupbearer—The Great Forfarshire Covers—Cullow Fair—Cortachy Castle—Highland Crosses—Laying down Permanent Pasture without a Crop—The Cortachy Herd and Dairy   .   .   . 211—230

CHAPTER XI.

CORTACHY TO PERTH.

Mr. Watson of Keillor—His Show-yard Career—Old Grannie—The Polled Herd Book—Experience of Southdowns—A Judge of Distance—Mr. Bowie's Herd—Dr. Murray of Carnoustie—The Story of The Cure—Major Douglas—Salmon Fishing on the Fife Coast—Speedie v. Seals—The Shark and her Whelps—Cock and Hen Salmon—The Carse of Gowrie—Rossie Priory—"The Scottish Henwife"   231—260

CHAPTER XII.

PERTH TO DUNKELD.

The late Lord Lynedoch—Net Fishing at the Lynn of Campsie—Mr. Speedie's Ice House—Salmon Prices—The Scone Steading—Mr. Paton's Gun Shop—The Caledonian Hunt Club at Perth—The Club Rules—Its different Places of Meeting—Forty-five Years of Racing
261—281

CHAPTER XIII.

DUNKELD TO BLAIR ATHOLE.

The Athole Larches—The late Duke of Athole—His Love of Otter Hunting—The Athole Forest—The Dunkeld Ayrshires—Milk Statistics—The Duchess Dowager's Farm at St. Colme's—Pitlochry—The Castle and Burial-place at Blair Athole—The West Highland Herd . . . . . 282—297

CHAPTER XIV.

PERTH TO KEIR.

The Strathallan Castle Herd—Walk to Rob Roy's Grave—Fern Thatching—The Braes of Balquhidder—Points of West Highland Cattle—Black-faced Sheep—"M'Claren's Cow"—Recollections of Deanstown—Keir—The Steading—Clydesdales, Shorthorns, and Garden 298—323

CHAPTER XV.
KEIR TO FIFE KENNELS.

Alloa—Mr. Mitchell's Herd—To Keavil—The Keavil Herd—A word with Mr. Easton—The Old Fife Breed of Cows—Fifeshire Feeders—Old Days of the Fife Hunt—A Visit to the Kennels . 324—349

CHAPTER XVI.
ALLOA TO SKYE.

The Wood of Caledon Bulls—The Duke of Montrose's Herd—Sail to Skye—Products of Portree—Ride to Duntulm—Cattle and Sheep in Skye—Skye Terriers—A Pig Hunt—Symptoms of Falkirk—The Poltalloch Herd . . . . . . 350—373

ENGRAVINGS.

"WITH PUNCTUALITY AND DESPATCH". . . *Preface*

HUGH WATSON OF KEILLOR . . . . *Frontispiece*

PROFESSOR DICK . . . . . *Vignette*

HEAD OF DUNTROON . . . . . *Page* 372

FIELD AND FERN,
OR,
SCOTTISH
FLOCKS AND HERDS.

CHAPTER I.

ABERDEEN TO THE SHETLANDS.

"Merrily, merrily, goes the bark
On a breeze to the Northward free;
So shoots through the morning sky the lark,
Or the swan through the summer sea."
SCOTT.

Departure from Aberdeen—Voyage—The Shetlands—A Lively Passenger—Cabin Councils—Arrival at Lerwick—A Peep at the Town—Walk to Voe—The Lonely Muirs—The Cows—The Ponies—Pony Life in the Coal Mines—The Sheep—Wool Manufactures—Improved Stock in the Shetlands—Back to Kirkwall.

Mc COMBIE, that standing menace to the vegetarians, bade us good luck on the pier. In the eyes of Captain Parrott and his men, the sea was of course "as still as a duck-pond"; but when we had passed the "Granite" and the "Heather Bell," and were fairly outside the bar, the prow of the "Vanguard" began to dip ominously, and the ground-swell told its tale in the bay. Oblivion seize those hours! Prostrate forms soon peopled all the couches and every inch of available carpet in the cabin. One voice of the night put in its feeble protest against our "using a head for a football"; another groaned piteously when its owner was roused at Wick and told that he was resting over the mail-bags in the floor; and a third had the presence of mind to suggest whisky at 3 a.m. for "buckling on a gentleman's boots."

Wick was fast asleep, but its lightermen knew no slumber, while we give out flour and meal from the hold enough for a beleaguered city. A navy of tan or chocolate-sails studded the offing. One by one they came slowly into harbour, some with hardly the tenth of a cran, or only a cod-fish to mock their toil, and others with their richly-laden net-tresses glistening in the moonlight, like a sheet of molten silver. Four hours more, and we are at the entrance of Kirkwall Bay, and passing the long, low island of Hellersay. A heron is perched on the ruins of its Picts House, and here too is Thor, the Shetland bull, whose namesake was supposed to fish with a bull's-head bait for the Sea Serpent. There is quite a jubilee on the Kirkwall-pier, while the packet is unlading; but the thought of thirty-odd leagues between us and Shetland sternly declines to be smothered.

Newspapers are a slender solace to begin with. We read languidly among the telegrams that The Ranger has won the Great Yorkshire, and Golden Pledge the Ebor; and, again, "it is four years since the game-bags have been so well filled." Then doubt is cast upon these returns by the Opposition. What are we to believe, when a *Banffshire* paper puts the Maharajah " Donald Singh" at the head of the list with 140 brace; and a *Perthshire* authority will have it that they were " principally cheepers," and that " the editor and head compositor could have done as good work with sticks"? They know no such joys and disputes where we are going. In fact, they have no papers; and the quiet natives pore with deep interest over the *Orcadian's* tale of the lapwing, which lost its young ones in a ditch, and died of grief on the spot. Now a new interest presents itself in the shape of a young English tourist, who is quite bomb-proof, and always exhorting the passengers to make an effort, and come upon deck to see something. *" Come up, indeed !"* as we told him—*" No, not to see the last of the water-bulls from the loch of Olginmore."*

The seas have been " restless" in these parts time out of mind, but they were nothing to him. In another quarter of an hour, he is among us again. He has conceived the idea of throwing a bottle overboard with a joint statement that we have been shipwrecked in a brig about this date, and are floating in Northern latitudes, with only biscuits on a spar. He will become a subscriber, he says, to the Orkney and Caithness papers for a quarter; and if they ever notice that bottle, which he is convinced they will, we can be communicated with at our respective addresses. One of the passengers consents to " sign anything for peace and quietness," but the rest spurn his pen and paper, and the project is ultimately quashed. His announcement of the Fair Isle makes him rather more popular, and a few of us peep feebly through the port-holes at those lonely fishers. They have no regular post, but the tacksmen bring them meal and take away fish; and so their life wears on. The captain chaffs them by putting on extra steam; and the game stern-chase which they row for the sake of the newspapers, which he opens and sends to them over the waves like flying-fish, would not displease Bob Chambers.*

A more protracted interval between his cabin missions is at last explained by the fact that our friend, in default of other excitement, has developed a tendency to sit on the bowsprit. In vain do the sailors warn him that he will become a prey to the lobsters; and of course a "difficulty" ensues with the mate, who follows up his "first warning" by collaring him. He argues the point vigorously with that officer, and fails to make any impression. The tea-party is any-

* A friend has furnished us with the following sketch of the Fair Islanders: "Between two and three years ago, the Fair Isle was found to be overpopulated, as the sustenance derived from the cultivated land was limited, and the fishing had for several years proved a failure. By means of a public subscription, a number of families were taken off the island, fitted out and sent to Glasgow, whence they were forwarded to Canada at the expense of the Emigration Commissioners. When these families passed South in the steamer, I made it a point to notice their development and character. I found the men—both young and old—stout and able-bodied, and apparently very intelligent; the women the same in every respect; and the children stout and healthy. They spoke very good English, and did not seem at all shy or put about when spoken to. In the summer time, I have seen several of the Fair Isle people at Kirkwall, which they reached in small boats, only fit to carry four men. On every occasion they quite answered the foregoing description. When they find their way South, they make first-rate sailors, and I have known them reach the position of commanders of large foreign vessels."

thing but sympathetic when he tells them how he has been " insulted" ; and when the captain appears in his place, and affects, with quite a dramatic start, to know happiness once more, after having been weighed down for hours by the report of his loss, he makes a final appeal. First he adduces precedents: "*I sat on the bowsprit between Edinburgh and Aberdeen.*" "*Did you?*" said the captain; "*but the sea isn't so deep in those parts.*" His great point is that he will lay the mate's conduct before the steamboat company; and there the captain is quite with him, as "*he'll get a gold medal for saving your life.*" But harmony is restored, and the evening closes with toddy.

It is nearly midnight, and we are at the "seagirdled peat-moss" at last. There are lights in every lattice when we steam into Lerwick Bay, and the "retaliatory" report of the cannon soon made us welcome. An August morning broke on a quaint, old, innless town, with one narrow, flagged street at the foot of a hill, up which you climb through alleys. They bear the proud names of "Pitt" and "Reform," but were more suggestive, during our brief visit, of perennial gutters, and washing-tubs. Dutch clogs, seal-skin purses, and comforters seemed great articles of trade, and there were also photographs of Earl Zetland, holding a grey horse. Job Marson and Voltigeur are names unknown. Commander Smith of the revenue cutter, and fully six-feet-two by sixteen stone, is evidently the great man of the place, and makes a brave figure as he goes in full uniform, with his cocked hat and clanking sword, to church. It is grand to watch his six blue-jackets pulling through the bay; and he had just derived no small lustre from having brought an American skipper most promptly to his bearings, when Jonathan declined either to hoist his bunting or to show his papers.

Most tourists take a pony and a guide, and sally forth after "thick little trout with red spots." Loch fishing, however, is not in our way; our toes would brush the heather; and Mr. Hay's Ordnance map, with green, red, and yellow lines, is an all-sufficient aid. The road out of Lerwick winds for a short space among boulders and peat-hags, then down a rocky path to the right, of which a woman with a huge caissie full of hosiery on her back seemed to make nothing, and over the strath at Dales Voe. The road became puzzling, so we asked a girl; but she took fright at the map, and hurried swiftly down the hill to a cottage, whose women inmates came out and surveyed us with as much zeal as if we were a travelling gorilla. So we fall back on our map, and leave the green line, which means "road prior to 1847," cut the yellow entirely, as it is merely "surveyed," take to the red "formed by the Relief Board" for a short space, and finally hit off the green triumphantly by Laxforth Voe and Wabister. Sheaves appear in a valley, with small farm-houses here and there, and black and white cows, looking no bigger than trick ponies, on the hills. The sides of the road are wild and open, and the cottars' white crops are rich and yellow—with runchy. Some few planes and stunted elders, and a hermit ash in a garden are the only apologies for a tree. At the distance is a lowly straw-thatched hamlet, and seawards a bold Bass Rock, which seems a very Gibraltar to the Cheviots.

Public-houses and milestones there are none. "*If you'll call at some respectable house, and say you're hungry, it's quite enough,*" was our primitive "office" on this head. Even among the humbler folk you can get eggs and oat-cake, and as much buttermilk as you can drink; and a trifle in the baby's hand at parting will send you away with a blessing. Eggs, it must be remembered, are fourpence a dozen, "except when the fleet comes," and then even the mildest Shetlander is not caught napping. For threepence you may generally buy a chicken, for sixpence to ninepence a good fat hen, and for eighteen-pence a goose. The latter are of the small grey sort, and are put out to wander on the hills from Whitsuntide to the end of November, and the procurator-fiscal descends with a month on any one who disregards the slit-mark in the web of the foot. A ewe costs about seven shillings; and a cottar's wife, who fastened her two-year-old stot with a cord and a peg among the ooze, and lamented that it was not "longer-leg-ged," thought it very odd of us that we would not close with her £2 offer. In short, for £60 one could stock a miniature farm, and have a team of four mare ponies. We are nearly a hundred miles from a grouse or a badger; but there are a few hares, and there never was more than one fox in the islands. He was an escaped Icelander, and a regular parishioner of Lunnesting, where he led a merry bachelor life among the rabbits while it lasted. He occasionally ran down the sheep, and, by way of comment on the mutton, only ate their tongues.

Gradually we begin to strike inland through a succession of rudely-fastened gates on to a boundless muir, which sheep and ponies seemed to hold in fee. A skewball and a brown with long matted locks stand moodily under a rock, one with a foal at her foot. A dun, as short on her legs as a Kerry, and a "cherry-red" chesnut trot off steadily to the heights along with a yearling, which winter will reduce into a mass of frosty wool, with fierce little eyes and four black feet. As for the sheep, they are off like a pistol-shot before one can say, "*Lie down, Croppies.*" A lusty lamb with Cheviot-looking wool, which will ere long run to hair at the points, scampers after the plainest of dams, which has milked till she is almost "at the lifting." Unity of colour there is none. One ewe was brindled, and others blue, grey, and black with white legs and blaze. "Black and all black" was quite common, and the colour would sometimes come rug-shape, or under the belly, or round both eyes. A pink or a white nose could hardly be called the exception. Still the palm for queerness rested with a lamb, speckled or ringstraked, we forget which, and with head and horns like the pictures of the Evil One.

As the day wore on, the silence would have been terrible even to a seasoned Crusoe. For two hours and a quarter we met no one save a minister, who asked for the news by the packet, and was told of the death of Lord Clyde. There was no bleat, no nicker in the drowsy distance, no cry of the curlew, no "wild birds gossiping overhead" in that peaty, treeless waste. A solitary rane goose was sailing in one of the vast chain of lochs, just tipped by the evening sun. The murmur of a little brook across the road was quite a joyful thing; and when we hadn't that solace, we wakened up the nearest sheep group with a "view halloo," just to keep life in us. Then our patience began to fail sorely. Why had we ever loved a shorthorn and nursed our love at Towneley and Athelstaneford? What had we done that we were paying this fearful penance by walking "eighteen miles on end" to an unknown Voe or a remorseless Pa-Pa? Shall we ever again join in those merry sales at Blenkiron's, and the annual call for Dundee? Why had we given up our herd rambles to English farm-houses, peeping out coyly among beehives and apple-blossoms, and redolent of fat bullocks and wedders, to roam in these ancient silences with three-year-old mutton at 4lbs. a quarter? Why, indeed! There was no lodge in this wilderness. Twice over we stretched away to what seemed one, at the turn of the road, but it faded into a greywacke rock.

The red line on the map knew no end. At last, we savagely doubted that map altogether, and longed for the gipsy patran. Half a mile more, and we hear voices—three girls at the cross-roads in their plaid gowns and crinoline complete. A house, a bay, and a smack at anchor!—the long-desired Voe at last!—

"The lighthouse top I see:
Is this the hill, is this the kirk—
Is this my ain countree?"

There was a kind welcome from Mr. Adey, and abundance of fresh materials for a morning's ramble. The kirkyard on the edge of the bay was a mass of nettles and sea-faring graves, in which rest "Lawrence Tait, mariner, and his spouse Lilœus"; and the list on the door disclosed the fact of only three voters in the lordship. Thousands of carcases of dried cod were piled in the Iceland smack, and a few dozen were swimming merrily in her well. The store near the landing-pier was boundless in its variety, and descended from whisky, food, and raiment to comfits and castor-oil. The ponies in the park were infinitely more sociable than their sisters on the muir, and an old dun mare merely put her lip down and her ears back, for conformity, while we wrote off our thoughts quite comfortably on her loins. That over, we strolled to the compound, and helping to "twist" two or three cows out of a lot for the South added a keen zest to breakfast.

The cows of Shetland are pretty much a pocket edition of the old-fashioned Yorkshire milch cow, but with more of the shorthorn head. They are principally black and white, red and white, black with mottle grey, red, dun, and "red mixed with ashes," which may answer to our dark roan. About Edinburgh, where a good many are kept to feed in parks, white and black, and more especially white with black ears, are the popular colours. If the former are crossed with a white bull, the calves in many instances fall rich black roan. The Unst cattle are best, but in some islands the bad-coloured ones seem only born to a course of seaweed and starvation. Queys in store condition, from three to five, average £3 to £3 5s., and stots of that age are scarcer, and from 5s. to 10s. dearer. Cows near calving range from £4 10s. to £5 10s., but unless they are very well fed the queys scarcely ever have a calf in Shetland till their fourth or fifth summer. Some cows are sold fat off the islands in August and September, and the fleshers deem an orange tint on the skin their highest test of quality. A few between seven and ten from the very best pasture have "died well" in Aberdeen at two to two and a-half cwt. neat. Others are exported younger, and are fed off, about Edinburgh, without ever having a calf; but the stots are liked better for stores, and, with good keep, will more than double their price in twelve months. Eight quarts of milk can be got at a meal, but five are not a bad average where no bean flour has been used. The bettermost class of Shetlanders have generally "butter and cream flying through the house;" but, except in Unst, there is very little cheese. The "bland" is made by pouring hot water on to butter-milk, and putting it into jars when the curds are removed; and at Voe they quaffed it like nectar, and vainly adjured us to do so.

The cattle are principally bred by ones and twos among the cottagers, and there are large gatherings of them at Lerwick on packet day, from August to October inclusive. Some years ago, when they were in less request, a friend of ours went over for some, and his wishes were made known, not from the pulpit, as has been sometimes the case, but by the beadle at the church-door. The news spread as quick as an Indian patty-cake, and nearly fifty were in the church-yard for inspection, when he looked out of the vicarage window next morning. "Proprietors' sales," as they are called, take place in May, when the small tenants sell their beasts to the dealer, on the understanding that he is to pay over the money to the proprietor, who thus secures his rent. They are all bought for grazing, and are put on some island when the grass is ready, to get freshened up for exportation later in the year.

There are a few garrons from the Orkneys for heavy draught; but nearly every one uses the ponies of the country. Duns are in great request; but the colour is not so much an object if the bone be only good. Greys and chesnuts are scarce: bay has not its wonted supremacy; and bays and blacks are most common. Some buyers began to go against piebalds, from a belief that they had Iceland blood in them, and were softer and slower in consequence. Five or six hundred of these Icelanders have been known to arrive annually at Granton, Aberdeen, and Grangemouth. They are, generally speaking, two hands higher than the Shetland ponies, and sometimes sadly stubborn on landing if they are not twitched under the lip.

The best ponies come from Unst; but both there and everywhere the breeders are far too indifferent to the points of a sire, as long as they are foalgetters. "About a quarter of Unst has a skeleton of red sand-stone and serpentine, with a thin soil studded with large red stones, and the knobs of rocks sticking up. Yet among these rocky incumbrances one sees scores of ponies picking the green grass, which the light of heaven and the breath of the Gulf stream force up from so barren-looking a bed. Still, Unst may be regarded as the heart of Shetland; and a sunny, genial-looking spot it is, when other parts of the country are dismal enough, in the late northern spring*." The heather and the bog-grasses elsewhere do not make much milk, and the mare ponies sink so much in condition that they are invariably barren every other year. If very well kept they reach 44 inches, but the average is from 38 to 42. Their owners frequently lose sight of them for a couple of summers, and recognize them when wanted, not by any formal "Exmoor brand" on the saddle-place or the hoof, but by a peculiar

* *Mark Lane Express*, Oct. 21, 1864.

slit or bits of tape, clout, or leather tied through a hole in the ear. Each cottar has generally a few ponies on the hill, and when the May and October sales at the different stations are at hand, they circumvent them for selection by the dealers with a line of forty or fifty fathoms. Still, the poor, hardworking Shetlander is generally little more than the nominal lord of his pony: poverty is his lot from the cradle to the grave, and, as the phrase goes, he is "still in tow." In his dire need the merchants become his mortgagees, just as the curers are to the herring fishers: they advance money on the security of his foal, and he doesn't get the best of it with "halvers' mares."

There is no need to call. "the oldest inhabitant" in Unst to witness that an Indian file of forty horse ponies has been seen there carrying peats. The Ashley Act has changed all that, and only left enough of them for sires in the island. In fact, such a demand sprang up at the collieries that the Shetlanders could not resist the lure of £5 10s., and "ground up their saplings" at two years old. Now the demand is less, and they are satisfied with £4 for them at that age. When the trade was at its height, upwards of five hundred were taken annually for the pits, and not thirty mares amongst them, and about two hundred for general use. They were of all ages from two to twelve, and for a very good one the pit owners would give the dealers as high as £8 to £10. The year 1857 was a red-letter one, and a noted dealer, Mr. Parris, of Kirton Mains, near Edinburgh, brought over as many as two hundred and twenty in two weeks, and four hundred in the course of the season. In 1861, no less than six hundred and sixty-six came South by the steamer, and perhaps fifty more by sailing vessels. Such heavy sales, which were continued in a modified degree for some years, nearly drained the Shetlands of aged ponies; but as the dealers' purchases have fallen off considerably these last two summers, the breeders have had a little breathing time. Now, a good horse pony and "a very extra mare" will average £7, and mares generally range about £2 below horses. The pit owners do not buy in December, as they are engaged balancing their books, but January and February bring a brisk demand, which dies out with the fires, and revives with them again.

The Welsh ponies outnumber the Shetland in the Durham collieries, and the Scotch have the lead in the Northumberland, where the present working seams are much thicker, and require larger ponies for "putting" purposes, or drawing coals from the "face," to the horse roads. For "putting," the pony height varies with the seam from nine to thirteen hands; but there are other "ponies" on the main ways fully fifteen hands by the standard. The Scotch ones (which are chiefly bred in Argyleshire, Mull, and Skye, and the western part of Ross-shire) average twelve two, the Iceland twelve, the Welsh eleven, and the Shetland ten. Ponies from five to seven years old are preferred, but nearly eighty per cent. are between two and three. The great majority are very tractable, and the most vicious recusants are to be found among the Welsh. Some of the ponies have not seen the light for fifteen years; and one horse at South Hetton descended in '45, and has not come up since. In well-regulated pits they are equal to well-kept hunters in point of muscle and condition. They have generally green food for a month during the summer; and at South Hetton and several other collieries, where Mr. Charles Hunting, V.S. (a well-known writer on the subject), is in charge, the oats, beans, and peas are crushed and mixed with bran, and the hay is always chopped. They suffer principally from indigestion, but not nearly to such an extent as agricultural horses, and scarcely ever from diseases of the lungs, glanders, or farcy; and if their eyes go, it is almost always from accidents in the dark. The runs vary from two hundred to six hundred yards; but the average day's work is twenty miles, half of it with empty tubs. One tub contains 10 cwt. of coal, and weighs nearly half as much again; and therefore when the seam dips five or six inches to the yard, the wear and tear of pony power is fearful. It is a perilous task, and broken backs, and necks, and legs swell the stable mortality bills to a very large amount.

As we have run the Shetland ponies right out of the islands to ground in Durham, we may go back and draw Unst once more for the native sheep, in which it and North Mavin are said to excel. They pass a strange, tameless existence, and continue to the last as shy as a rabbit. In June their owners muster a *posse comitatus*, of all ages and sizes, to sweep the hill. A dog is worse than useless, as the sheep never "pack" in a panic. A skilful woman can pull one in five minutes, and a three-shear (if we can use the term) will produce nearly 2lbs. There are three kinds of wool on one sheep, all varying in quality. The fine-woolled or "the beaver" sheep has this fur or down all over it, under the protection of the coarse hairs; whereas it is only found on the neck and a few other parts of a less kindly one. The white and light-grey wools vary from 1s. 8d. to 2s. 2d. per lb.; and brown, or a peculiar shade of it called "Mooriah Mound," will reach half-a-crown. The best sorts are knit by hand into those veils which defy Æolus, and those still more remarkable shawls from a yard to two yards square, which can be drawn through a wedding-ring, weigh little more than four ounces, take upwards of a year to make, and are sold as high as five guineas. Stockings are generally made from the coarser sorts, but a pair from the high-class wool will fetch a very great price. Every part of the staple has its use, and the refuse, when decomposed after the stocking process, is made into hats. If the sheep are taken south, they still carry the traces of their bleak and hedgeless birthplace in their manners and their blood. A seven-foot wall will not keep them in, and their storm-tried heads despise all shelter save the sky. So much for the native breeds.

On the farms of Vinsgarth, Reawick, Bigton, and Maryfield the shorthorn has surely made his way for some years past. The first and second cross heifers have been kept for cows, and the young cattle are generally sold as two-year-olds. The Angus bull has been used to some extent at Quendale; and Symbister can boast of a pretty good Ayrshire. Shetland, however, is perhaps rather better adapted for producing sheep than cattle, and where the commons have been divided the native breed has been crossed with the Cheviot, in a few instances to three or four generations. Still, the ewes and tups have been so often selected at haphazard that the offspring is a very indifferent sheep, which will hardly bear the expense of shipment to the southern market. The native ewes are now crossed much more frequently with a Leicester, and the produce are readily picked up by a certain class of buyers for the south. A flock of Cheviots has been kept for some time past by Mr. Bruce at Vinsgarth, and part of them have been crossed with an Oxford Down from the Duke of Marlborough's flock; while Mr. Walker, of Maryfield, has used a Southdown tup. Pure Cheviot ewes are also kept at Lunna, Laxfirth, and Bigton, for breeding half-bred Leicester. The proprietors in Hoy, Noss, Fetlar, Sumburgh, Gremista, Hillsburgh, and Vementry have more or less of the Cheviot blood, and are gradually progressing.

Still, the difficulties and expense of obtaining and upholding a pure Cheviot flock in Shetland are very considerable. The moist climate, undrained pastures, want of enclosures, and consequent lack of shelter induce braxy to a great extent, and hence the flockmasters are anxious to part with their lambs at any price, rather than lose them altogether in the winter. The consequence is that, with few exceptions, the Shetland sheep stocks are of a very mixed character. Tenants are decidedly anxious and willing, but their aspirations are very much in advance of the encouragement given by the proprietors.

The first cattle show was held at Lerwick in August, '64. Entries came from all parts, some by very indifferent and rugged roads, while others had of course to be put into boats, and ferried over the voes. The Highland Society gave prizes for cattle of any other pure breed than Shetland, and Aberdeenshire and Caithness furnished the judges. In the improved cattle and sheep classes the prizes were awarded almost entirely to Walker of Maryfield, Bruce of Vinsgarth, and Umphray of Reawick. The entries were generally pretty good; but the pig part of the show dwindled down to a brace of sows, and a boar whose prize was withheld. The "old original" cattle and ponies of Shetland mustered well; but to bring the sheep, or to ask them to stop when they were there, was far beyond the power of man. As well try to lot and sell by auction white bulls from Chillingham Park, or the old Forest of Caledon.

One of those hopeless afternoons "which wets the puir Scotchman to his sark, the Englishman to his skin" did not improve the aspect of the "scattalds" or undivided commons, as we trudged back from Voe long before the signal-gun boomed out its half-hour warning over Bressay. The deck was quite a Shetland cattle market, and it was elysium to be once more among the busy band of farmers giving orders about their stock and getting a few last words with the captain. The reports of our mercurial friend were conflicting, but on the whole favourable. He had rung his bell at short intervals all the first night, and got up rather low-spirited on the morrow, but had ultimately gone shares in a pony gig, and departed into the interior with a cattle dealer, who was anxious to show him life. The lights of Lerwick were soon far on our lee. Once more stretched in the stern, and with nearly "forty miles in us," we revisit the Fair Isle only in dreams. The sun is up and bright when we reach Kirkwall, and the Shapinsey mail is cleaving her way through the long seaweed tangle.

CHAPTER II.

THE ORKNEYS TO THURSO.

"My horses were in good condition. Dandy and Billy, the coach-horses, were as sleek as seals. Gentleman Dick, my saddle-horse, showed manifest pleasure at seeing me, put his cheek against mine, laid his head on my shoulder, and would have nibbled at my ear had I permitted it. One of my Chinese geese was sitting on eggs; the rest were sailing like frigates in a pond, with a whole fleet of white-topknot ducks. The hens were vying with each other which could bring out the earliest brood of chickens. Taffy and Tony, two pet dogs of a dandy race, kept more for show than use, received me with well-bred though rather cool civility; while my little terrier slut, Ginger, bounded about me almost crazy with delight, having five little Gingers toddling at her heels, with which she had enriched me during my absence. I forbear to say anything about my cows, my Durham heifer, or my pigeons, having gone as far with these rural matters as may be agreeable."
WASHINGTON IRVING.

Orkney Nicknames—Shapinsey—Defence of Hellersay—Orkney Panorama—Mr. Balfour's Shorthorns and Shetland Ponies—The Legend of "Spunky"—Orkney Sheep Crosses—Sheep Marks—Devon and West Highland Bulls—"Fishing Pork"—Orkney Garrons—The Fortescue Harriers—Swanbinster—Postal Difficulties—A Sea Sick Horse—The Message from the Ice—The Hoy Farmer Consoled.

"SANDEY and Burray for rabbits, Rousay for grouse, and Shapinsey, Stronsey, Westrey, and Sandey for crops and cattle," was the terse synopsis of the Orkneys by "a friend in council." Then he waxed more diffusive, and told off on his fingers some of the fancy names of the natives. The men of Hoy are "Hawks," because they once "supplied the great falcon from their cliffs to the pageants of crowned heads." The "Seals" dwell in North Ronaldsay, and the "Awks" (a diving sea-bird) in Westrey.

Rousay men will be "Mares" to the end of the chapter, because their learned legates quite overlooked the necessity of bringing back a sire, when they executed their horse commission upon the main land. The natural bias of its sons towards the rich soil of Stronsey is aptly typified by "Limpets." Harra does not touch the coast, and therefore "Let be for let be," as the Harra man said to the crab, when he clutched it in his first wanderings by "the sad sea waves," has constituted them " Crabs" for all time. It is in these epithets that the Orcadians, if they have a difference, hurl their mutual scorn, and a bloody nose is sometimes the sequel. The Shapinsey " Sheep" are just as touchy as any of them. On one occasion, when they were cutting peats in a thick fog on the Foot, they were assailed with *" Baa, Baa,"* from a passing boat. Flinging their spades and tuskars aside, they pursued the aggressors in a boat, with threats of condign vengeance, half way to Stronsey, and then found that they were only sheep after all.

The eye of Washington Irving never rested on the maroon and green velvet of the Orkney caves; but Shapinsey, which his parents quitted six months before his birth, was the home of his kith and kin for many generations. This island is six miles by three, and the sole property of Mr. Balfour. With the exception of a few primitive patches of grass and heather, it is now all reclaimed. The acreage under plough has increased from seven hundred in 1848 to nearly six thousand in '63; and Mr. Balfour has also a large number of tenants in the adjacent islands, which he visits twice or three times a year in his yacht, with Marcus Calder, his factor. Some of the Shapinsey holdings, which are let on improvement rentals to begin with, range from one or two hundred acres down to twenty-five, but many are limited to one of the ten-acre fields into which the island is parcelled. The fence-lines are drawn with mathematical regularity, and every farm has a name, be it Quholm, Inkermann, Balaklava, Lucknow, Ganderbreck, or Bashan. The Orcadians, except on provocation aforesaid, are a peculiarly quiet race, and no oath, blow, or drunkenness has ever been known in the Shapinsey revels, which principally resolve themselves into Highland games, sack races, and rifle-shooting. They are also of a highly-studious turn, and very fond of astronomy; and as for "Allison's History of Europe" they have fairly read it to tatters.

Gudin delighted to paint the deep, yellow sunsets; and strangers and even natives are restless under the great length of Midsummer twilight. Mr. Balfour takes it calmly enough in those lovely "Lapland nights," and has often needed no reading-lamp near a south window, at two o'clock a.m. Vegetation is seldom at rest, and the yellow jessamines for the Christmas decorations of '59 were plucked in the open air. The grass season is "longer than a Syrian harvest;" a fifth crop was once cut late in December, and the pasture had made good head by Twelfthday. Winter comes with the nip of March, and hence the hardy trees which make their effort before May have invariably failed. The white-skinned and the mountain ash, the alder, the bayleaved willow, the plane, and (where the land is deep) the elm, have all struggled through, but it has been with pain and sorrow. The varied music, which connoisseurs profess to find in their rustle, is drowned in one thorough bass, when the west wind sweeps the chords, and "shaves the twigs as with a bill-hook." Where the vapour of the sea-spray floats through the air, its presence is marked by the redness of the larches and the greenness of the pastures, which are said to find in it an antidote to sheep-rot. The grass of the South isles is not equal to that of the North; but the hay season is pretty universal in July and August, and the harvest-homes at the end of October usher in a long, calm back-end, or " Peerie Summer." In old days, black oats or grey followed bere, varied occasionally with potatoes, and then white gowans and weeds to rest; but modern Orkney farmers are wide awake. The fiveshift is pretty universal, and wheat can be got to 63lbs. and 44 bushels to the acre. Perhaps the most tenacious heritor is the ox-eyed daisy, which defies eradication, and sometimes covers a field like snow.

There are two Markets in Shapinsey, summer and winter, and the Agricultural Club meets on the school brae to give prizes for stock, poultry, butter, cheese, eggs, &c., in August ; and in February for a ploughing competition. Mr. Balfour soon extirpated the original breed of sheep as being utterly worthless, with the exception of a few which have been crossed with Southdowns for home consumption. The sheep now on the island are Leicester-Cheviot or "halfbreds," as they are strictly called all over Scotland, to distinguish them from crosses. The cattle are crossed with Shorthorn bulls, the pigs with the Buccleuch breed, and the garrons with Clydesdales, whose fifteen two and three descendants are gradually supplanting the pony teams. The Balfour Castle herd began with pure Shorthorns from Chrisp and the Brothers Cruickshank. Females of nearly "Herd-Book" blood, were added from Sir John Sinclair of Barrock, Sir George Dunbar, and others; and crossed chiefly with bulls from Sittyton, Dishforth, and Kingcausie. West Highland bulls have been furnished to some of the smaller tenants, in order to establish a more thrifty feeding race. Shetland queys have also been purchased, for the sake of improving the size of their stock, by better keep and careful selection for crossing, as well as the presumedly enduring impress of a first impregnation by a shorthorn.

Strolling down the cunningly-receding walk from the castle, and so through the old stone gateway, with its grotesque carvings of syrens, satyrs in knickerbockers, and cats with bagpipes, we reach the shorthorns in Stronberry. Cruickshank's Artisan draws himself up on a small knoll, among the calves, hard by a whole Mason tribe from Gulnare. The old red cow is there, with the roan Queen of Cruickshank's Empress descent, and great in the milking vein; and the udder of a beautiful black and white Shetland dame prophesies still more decisively of Scotch pints to come. The starlings rise in a cloud as we scale the stile, and stand among the Timothy-grass on the farm of Agricola. Beneath us is Kaiserklett, or Cæsar's Rock, where, according to tradition, Agricola's trireme was wrecked, and left its only sign in the hilt of a dagger with Neptune's figure struggling through verdigris. But there is a more peaceful tie with the past. A Pict's house was found in what was once a small loch, but where oats are now waving, on the top of Nearhouse Field; and the very next year after it had been drained, two plots of wild mustard, the infallible sign of old cultivated ground, sprang up on each side of the ruins.

Half the North isles of Orkney lie in a map before us: the dark heath of Edey, the rich pastures of Phurey and Westrey, Egilsey, with its old Pictish tower, the scene of the martyrdom of St. Magnus; the thriving Rousay, once the home of wild hogs; and the hill of Gairsay, whose barley has not lost its good name since Swein Asleifson brewed his own ale for those cruises to Cornwall and the Northumbrian coast, which filled *Orkneyinga Saga* with his exploits. The remains of the castle of Kolbein Hruga, a rival Viking, are still to be seen on Wyre, the low, green island at the foot of Rousay, whose fine western headland of Skeabra stretches boldly towards Eynhalga, or Holy Island. This earliest home of Christianity in the Orkneys still lends its name to the beautiful strait that divides Rousay from Evie and Rendall, which are full of small fields of arbitrary rotations, like a Dutch concert, of all shades of green. Cottiscarth, Binscarth, and the dark grouse hills round them are memorable in more modern times. Even a Cabinet Minister girded up his loins and fled from them at the news of a summons for poaching; and it was there that the officers of the fleet roamed lawlessly, five-and-twenty abreast, in search of starlings and birds of warren.

There, too, on the more lowly left, are Kirkwall, and its cathedral sacred to St. Magnus, the Mull of Deerness, and Hellersay still guarded by "Thor" with such fidelity and valour. The bit of red bunting hangs listlessly to the mast, as the becalmed letter-packet, inch by inch, drifts home on its evening trip. The *Streamlet* yacht lies daintily in the Bay of Elwick, where Haco once mustered his fleet, and where a seagull is stepping to meet the tide, with his breast and his toes out, as stately as a Chancellor on the first day of term. The eider-duck haunts these shores; the tern comes like a spectre in the fog, and lays its three blue eggs in the grass; and the otter is true to a home, whose family crest is the head of Ottar the Dane, and where its form is carved in stone, as a token, over the castle-gate. Night after night, this dark tan fisher may be seen sliding noiselessly into the creek below the volunteer batteries, unvexed with thoughts of Dandy, hound, or gun. Shapinsey gunnery has a far nobler object. Ninety charges were dealt out to the volunteer artillery troop, and they smashed the target fifteen times on three practice-days, thirteen hundred yards out at sea. Not a few of Washington Irving's kinsfolk were among the eighty which signed the Balfour muster-roll, and fell into rank behind the guns, and one of them still occupies the old house at Quholm. Inspired by the sight, we formed forthwith part of a storming party to Hellersay. Marcus Calder, that epicure in walking-sticks, fitted us all round from his armoury, the aggregate result of his visits to too confiding friends. Title or no title, they were all needed now. Thor steals down with a roan Sarah and a red Hagar to the beach, to dispute our landing, and manfully lows defiance, up to his very knees in the sedge. Vain is the form of Volunteer Allegiance in the provost of Kirkwall's pocket. Thor has defied Professor Aytoun, sheriff of Orkney and the Shetlands from the selfsame spot before, and what cares he for a provost? Still he retreats under the stick-and-stone practice, kneeling as he goes, and whirling the black loam far above his head, but only to advance, on our departure, with a louder flourish of trumpets than ever.

The Druid heads the Shetland pony contingent on Hellersay. All the foals are by him; and among his dun, brown, and mealy-grey dames are Lady Grey and the aged piebald Lochelia, with her middle close to the ground. Duke, Jolly, Duchess, and Barney are not in the troop. Those pretty brown pairs do the phaeton work; and the run from Worcester to Malvern, nine miles in the hour, is quite within their scope. Colonel Balfour, grandfather to the present proprietor of Shapinsey, began pony breeding at the end of the last century. He improved the form; and when the colours did not come as they expected, the natives, with a few drops of whisky to quicken them, laid the entire blame on Spunky, the Orcadian water kelpie. He was black, say they, and the sire of some of the finest original ponies of the islands; and if he was disturbed in his courtships, he vanished under the waves in a mass of blue flame. The Hellersay stock have been quite able to dispense with him, as North Unst has furnished them with some of its choicest jewels. Brisk, the chesnut, dates very far back, and headed the Balfour stud for well nigh thirty years, and his brother Swift was in the flesh for nearly forty-six.

The piebald Cameron cost £24, and although he rather spoilt the colours, he introduced a better shape, a smaller head, and decidedly truer action. Odin, of the same colour, also kept up the form; Thor got them nearly all skewballs like himself; and Lord Minimus was a grey, and the sire of grey beauties. They are shifted from island to island as the grass suits, and require the most careful drafting to keep them at nine hands. Mr. Balfour has about forty in all, of which the majority are duns and creams; and they are always broken at three, and made very tractable in a week. Her Majesty has had a pair of them; and some of the more fancy colours were once picked up by Ducrow.

Hellersay is also held by some crosses of Southdowns with the native Orkney, which have horns and tufts, and are nearly as bizarre in their shape and colour, but weigh a trifle more. The second cross with the Leicester tup is more delicate, but on good pasture both first and second average from 18lbs. to 20lbs. per quarter. In shape, but not in size, the first cross strains to the dam; and when the Orkney tup is used, the lamb still keeps the "bristling" head and scrubby tail of its sire. On the whole, the Southdown cross seems to "nick" well, but the twist is hardly full enough till the third generation. The native sheep have always been a vagrant race, and they follow the tide when it ebbs, with a fine eye to the seaweed. A Highland Society essayist speaks of them forty years ago as only clipping 1½lbs., and "thriving in holms not secure from eagles and corbies." In Shapinsey, about that date, there would be fully fourteen hundred, and men who never paid a half-penny of rent would have flocks of sixty or seventy. In Holm alone there were fully nine hundred, where none are to be found now, as the cottars will not use a tup of the sort, and go for a cross of

Leicester or Southdown in their keery. At "sheep-run day" the owners used to meet, with whisky, cheese, and bread, on the hills or shore, and examine the marks before clipping. Petticoats, frocks, and blankets for home use were all made of the fleece, as well as a favourite black-and-white serge, which was once much worn in Burray. The marks would have puzzled a weird to decipher, as on the main land alone there were nearly a thousand. The "fordren elm" mark was a piece out of the fore part of the ear, and the slit and hole variations on it were legion. Rigid laws were in force about overmarking; and Orcadian invention had to exhaust itself on rags of many colours sewed into different portions of the wool. One proprietor gave the cottars three years' notice to claim and take away their sheep, and then he made a clean sweep of the marks by simply cutting the ears off. The days of such communism are over now. At first the Orcadians were quite moved about their "rights," and it was only when their sheep had been seized, and a few Dutch auctions had come off, that they bowed their heads to a more rational *régime*. Mr. Archer Fortescue was very decisive on the Orkney mainland, and well supported by the other landlords; and we remember how, on our first walk from Stromness to Swanbinster, one of these ex-flockmasters described him as quite "a man of wrath," and conjured us not to set foot on his hill land, even if we did save a mile thereby.

Mr. Malcolm Laing brought in the Merinoes more

than half-a-century ago, as well as pure Cheviots from Roxburghshire, but they all dwindled away. Then Mr. Heddle of Hoy, Mr. Fortescue and some others began them again. Not a few came from Mr. Gunn; and for some years past the farmers have paid £5 to £8 to Caithness breeders for Leicesters, and generally go in for half-bred lambs from Cheviot ewes of their own rearing. Mr. Learmont, of Housby in Stronsey, is the only one who puts the Leicester to half-bred ewes; he also uses half-bred rams, to prevent the flock from getting too fine; and Mr. M'Kenzie, of Stove in Sandey, has both Southdowns and Shrops.

Cattle, store and fat, of all ages from yearlings to threes, are exported for nine months of the year to Aberdeen, Edinburgh, and Banffshire; and yet the home market is so well supplied, that when the Channel Fleet called at Kirkwall 6,000 lbs. were furnished daily during its stay, and ten days' rations to boot, "without deranging a single horn." Such a thing as a pure Orkney bull is hardly to be found. They were larger than the Shetland, generally black, with a white stripe down a razor back, and drooping hind quarters. Since the beginning of the century there have been West Highland bulls from Dunrobin, which suited the Orkney cows; but after the first cross the milk rather fell off. Devon bulls were another introduction of Mr. Malcolm Laing's, and the cross fed kindly, but lacked under hair, and looked rather more critically at the heather than the

Highlanders. The Devon bull held his colour when put to rough Angus cows, and the stock gained in maturity what they lost in hardihood; but still they were coarse in their points, and the cross was pretty generally considered "too sharp." The calves by a Devon from a half shorthorn Rousay quey came shorthorn in colour and character; and from a half Orkney and West Highland quey they were red and Devon-headed. Mr. Fortescue brought over Barometer and some other Angus bulls from the Portlethen herd, but the second cross with the country cattle did not pay so well as the first. A Hereford bull has been recently imported by Mr. Cromarty, of South Ronaldsay, and is the first of the "red with white facings" that ever reached these shores.

In 1840, only two or three Orcadian farmers had shorthorn crosses, whereas now considerably more than two-thirds of the cattle are so bred. At that date, there were only a few West Highlanders in Burray and Holm, and the rest were Orkney blacks. So great has been the progress, that tenants can now sell young beasts at £14 where £3 10s. was then thought a catch. Mr. Bakie of Tankerness, Mr. Trail of Woodwick, and Mr. Petrie of Graham's Ha', began the shorthorn bull system; and since then all prices from £50 to £20 have been paid for them to Yorkshire and Aberdeenshire breeders. Very few pure shorthorn cows have been imported except by Mr. Balfour, and he and Mr. Cromarty are the only breeders of bulls. The foundation on which the farmers have had to work is a very cross-bred one; and the length of leg and lack of hair, which are too manifest in many of the stock, confirm the truth of Mr. Scarth's opinion, that the West Highland bull must be used once more as a corrective. The calves are generally dropped between February and April, and run with their dams till nearly harvest-time, when the men and women have no English thirst for beer, but are quite satisfied with two quarts of milk a day.

The native pig is long-legged, coarse, and hard to feed; but the Buccleuch breed and the improved Neapolitans from Sir George Dunbar have wrought a healthy change. Except in one little island, the farmers only keep pigs for their own use, and much of the pork, which is sold and shipped at Kirkwall, is chiefly fed for eight to twelve months by the fishermen and cottars. The "fishing pork" has always a high yellow colour, when salted, which is not to be wondered at, as the contents of the trough are generally fish offal, "tang"—which grows, like bent, high on the rocks—sea-weed, and turnips, all boiled together. The very eggs of the hens, which follow the ebb and eat insects among the rocks, catch a peculiar flavour; but the hen-wives say, "*It's nae odds—it's only an egg for sale.*"

Very little attention is paid to horse-breeding. When Orcadian garrons *were* garrons with a mould of their own, the colt used to live under the cottar's roof, and have a green sheaf thrown to it through the large aperture, which made it a joint-tenant in the fire-heat; but this Arab-like feeling has worn itself out. Mongrels from Caithness have ruined the stamp, and the farmers now cross up their small mares with Clydesdales or whatever comes first to hand. The grey Sunbeam brought in some blood, and made matters better for a time; but, as an old farmer said to us in the *Vanguard,* "*All this wild crossing doesn't do; we'll never build up our garrons again. We attend far too little to it. The mares are not the biggest of the two, and we're quite wrong there.*" Hence the game Orkney garron, pure and simple, with his strong fore-end, straight hind-legs, and good couplings, lives pretty nearly in memory. Prices for horses are wretched, and fully 120 per cent. less than they were during the Russian war.

Still those Orcadian sportsmen "who followed Jehu," as Mr. Fortescue was termed when he first introduced them to the "merry harriers," are generally very fairly mounted. Some of them ride furiously, and others are found consistently on eminences, "enjoying the sport and the scenery." The cottars had never seen a leap taken before this new era set in, and were nearly as excited thereat as an old woman in one of the Border counties, who rushed forth and clasped her hands frantically, when a scarlet went in and out of her potato-garden— "*Niver let me see that bonny young man kill hissel.*"

The hunting does not begin until Mr. Fortescue's return from Aberdeenshire, late in January; but the pack are generally equal to forty or fifty brace, hunting five days a fortnight up to May morning. The kennel at Swanbinster is near the sea, and more than once the deep toll of some of the Hebden blood has acted like an Inchcape bell, and prevented sailors as well as the master of the pack from running in a fog on that treacherous shore. Dick Smith, the first whip and kennel huntsman, is a great character, and enthusiast as well. He was born, like his master, in Devonshire; and his hunting budget of " Horner Wood, Withy Pool, Winnesford, and all that way," is inexhaustible. As for the story of the stag which took the sea near Porlock it is quite an epic poem, when told with his curious chuckle. He has one very cherished link with Dulverton in " Crowner," whose grandam, as he duly impressed on us, was from there. The Huxwell drafts have done good service, and so have the Eamont and the Holker. Eamont Bluecap was quite a pilot till Gipsey arrived from Wales; Bachelor has a strain of the bloodhound in him, and "knows to a nicety when she squats"; Bustler is quite a guide-post on the roads; and Royal swings himself round in his cast quicker than any of the ten couple. Dick grows vastly excited when she "begins to lollop," and not only rides hard, but strictly for the pot, of which, strange as it may seem, on one occasion a pig partially deprived him. The hound Monitor is his " difficulty" : he *will* call him " Wallater"; "*for, Maister, I never could remember the name of that theer hound.*"

Mr. Fortescue has 3,500 acres of his own to hunt

over, of which a sixth is enclosed, drained, and under rotation, and the rest hill and bog. The whole was purchased eighteen years ago at about 45s. an acre. The house and steading are at Swanbinster, rather in a hollow near the sea, half-way between Kirkwall and Stromness, whose fish offal, ashes, &c., are thus very handy for manure. There is not enough sun for mangolds, but carrots grow well, and so do Swedes and Aberdeenshire purple tops, and green-top yellows. The large hay colls which are built up round poles, and tied down with cords for a month before they are put into the rick, are quite an improvement on the Cumberland "pikes," and everything about the place gave evidence of vigorous management, and wool at half-a-crown a pound. The past only peeped out in the little grey bridge to which the honeysuckle was clinging, and the plot of purple heather which was kept intact to mark the victory of the tile and the ploughshare. Mr. Fortescue has both a Cheviot and a half-bred side of the hill, and the older half of his thirty score of Cheviot ewes are all put to the Leicester. Some of his tups are of three parts Sanday blood, and he has also introduced a Lincoln to Orkney. The herd comprises fifteen Angus cows, and he intends to keep them pure, and only put his "shot queys" to a shorthorn.

The sloop was lying at the pier, waiting for a breeze to start with the draft Cheviot ewes to Aberdeen; and if the packet is indisposed during the winter months, she sometimes takes the mail-bags across the Pentland. It is, however, only of later years that there has been a regular winter post. In 1848 the local inspector for the Drainage Loan Commissioners found three official letters in his bag one morning. No. 1 directed a survey and a report; No. 2 asked why it was not sent in; and No. 3 was in the shape of a sort of peremptory mandamus to him, to defer no longer answering No. 1 and No. 2. The reply was to the effect that winter correspondence with the Orkneys always required a six weeks' margin, seeing that they were sometimes twenty-one days without a mail. During the summer, letters posted in London on Saturday night are nearly always delivered at Kirkwall on Tuesday; but in winter there are still terrible correspondence gaps.

From Bowscarth to Stromness was a weary beginning on the saddle; and there was a most vivid realization of the Orkney phrase, "he blaws and she wettens." When we did get there, under every discouragement, considerably before the time appointed for shipment, we were told on the pier that they had *" been waiting half an hour for a fellow with a horse."* Well might a fellow-passenger subsequently confide to us, on fresh provocation, *" They don't keep no clocks here—they ought to be put in the papers."* Stromness, with its lonely ash-tree and its zig-zag outline of house and jetty, looked like a city of the dead, as dawn grew into day. A strange steamer was moored in the harbour, and as quick as thought its boat was lowered and at our side. " *Re-*

port at Lloyd's, Award and Sultan lost in the ice—crews saved," said the fur-capped captain; and off it shot on its shadowy track. The whole scene read like a passage from "The Phantom Ship."

Our mate was quite learned on the subject of sea-sickness in animals, and esteemed pigs happy because they can get relief. There is one advantage in looking after a very sick horse for three hours and a-half—which we spent in actively checkmating its efforts to back down the hatchway, or make a bone-mill of the engine-room—that we had no time to be sick ourselves, and rode it out like a sea-gull.

We were soon past the Orkney Lighthouse, and labouring along two miles an hour against a heavy tide—

"Where hawk and osprey scream for joy
Over the beetling cliffs of Hoy."

The lyre birds, which are said to be great appetizers before dinner, and so fat, that with a wick through them they can do good candle-duty, haunt that "Old Man," who has had his share in many a hecatomb of victims. Still the survivors, to quote the words of a small farmer, when he requested Lord Macaulay's uncle to marry him, would seem to have found consolation: "*Oh! sir, but the ways of Providence are wonderful! I thocht I had met with a sair misfortune, when I lost baith my coo and my wife at aince over the cliff, twa months sin; but I gaed over to Graimsay, and I hae gotten a far better coo and a far bonnier wife.*"

CHAPTER III.
THURSO TO HELMSDALE.

"He had ridden towards merrie Carlisle,
　　When Pentecost was o'er:
He journeyed like errant knight the while,
And sweetly the summer sun did smile
　　On mountain, moss, and moor."　　SCOTT.

The late Sir John Sinclair—Caithness Sheep Farming—St. Mary's Mass—Georgemas Tryst—Shorthorn Crosses—Barrogill Castle—Barrock Plantations—Shorthorn and Galloway Crosses—Bringing-up of Calves and Yearlings—Sir George Dunbar's Leicester Flock—Mail Journey along the Coast—A Night Ride on Horseback.

THE mare was fully two stone below her Orkney form, when we saddled her once more, and led her off the packet at Scrabster Roads; but a few hours of stable-quiet brought her round, and we were soon jogging leisurely past Thurso Castle. Sir George Sinclair does not farm his estate, but is content with the heritage of a great name, which meets you in every page of the early history of the Highland Society, and which will live as long as a Cheviot ewe crops the heather of Langwell. A ballad is never really popular till it is whistled at the plough or set to the barrel organ; and we overheard the quaintest Lowland appreciation of that unresting baronet's labours in the denial of another tumbler of toddy to a toper, unless he could say, *" Sir John*

Sinclair's Statistical Account of Scotland," with proper emphasis and correctness.

If "the living serpent went forth" from Caithness "to carry its youth and vigour into other lands," it left something more than its cast-off skin to the county. The open winter, combined with good grass and turnips, has made it quite a rich storehouse of shorthorn crosses, and big, fine-woolled half-breds for the feeders and breeders in the south; and a Howard steam plough is working well at Barrogill Castle, within a few miles of John o'Groat's. Mr. Smith, of Olrig, has a flock of "Shrops," and crosses them for early lambs with his half-bred ewes; but with this exception, the sheep-farming of the arable part of the county may be said to consist in breeding half-breds, and turniping Cheviot hoggs from its mountain ranges or from Sutherlandshire. Leicesters came in with the late Mr. William Horne of Scouthal, more than forty years ago; but the want of draining and sufficient enclosures was against them.

We found their first traces on the Crown lands at Scrabster, which are held by Mr. Hay. Sir George Sinclair's farm, which is close to Thurso, carries a large number of half-breds, and its present tenant, Mr. Donald M'Kay, also rents other farms, for turniping the Cheviot hoggs of his Skelpick flock. Another well-known Sutherlandshire flockmaster, Mr. Patterson, occupies the Rattar Farm of the late "Shirra Traill," in connection with his Bighouse

Farm, which was once held by a son of the celebrated Patrick Sellar, and extends some twenty by eight miles from the west coast to Strath Hallidale. The late Mr. Gunn, of Glendhu, pursued the same plan at Greenland, a farm of Mr. Traill's, who has been the knight of the shire for nearly a third of a century, and whom we blessed most heartily for a little shade, as we made the turn near Castle Hill. Weary as we were of the sea, there was no resisting the graceful wave curl of the Dunnet Head, on whose sandy links, St. Mary's Mass, once the great fair for Orkney cows, woollens, and garrons, is still held in August. There was a day when its garron average touched £14; but even "Swifty Stewart" is not found there now, and it is at Georgemas, Kildary Great Fair, and Bonar Bridge that he goes in for the colts and the horse ponies under fourteen hands, which he takes down thrice a year to his Banffshire and Morayshire customers.

At the Georgemas tryst in 1863-64 Mr. Traill made the highest figure for his half-bred hoggs. The scene both of it and the Caithness county show is a piece of enclosed hill ground five miles from Thurso, on the high road to Wick, and the tryst is held on the Tuesday before the Character Fair at Inverness. Buyers of cattle, sheep, and wool crowd the hill that day from every part of Scotland, and the North of England as well; and since their secretary, Mr. Geddes, of Orbliston, bore his testimony so decidedly before the Morayshire Farmers' Club as to the superior length, strength, and hardihood of the Caithness gimmers, the breeders have begun to separate them from the wedder hoggs. The sheep number between seven and eight thousand; and on the last occasion the Castle Hill half-bred hoggs were quoted at 42s., and the half-bred lambs from Latheron Wheel at 22s. 6d. Mr. Traill buys his lambs; but the great majority of the farmers breed their own, and sell them fat as hoggs off turnips (folded or on the grass) in the spring, or in good store condition "on the hill."

Major Horne of Stirkoke is a very large breeder of half-breds; and it was by his kinsman, the late Mr. William Horne, of Scouthal, who was urged on by the counsels and example of Mr. Rennie, of Phantassie, that shorthorn bulls were introduced upwards of forty years ago, to cross the West Highland, or the large red or black, with brown tinge cows, which seem to have existed for ages in Caithness. "Shirra Trail," who, like his coeval, Lord Duffus, (then known as Sir Benjamin Dunbar), will always be remembered as one of the veritable old Caithness worthies, was equally zealous in the cause, and so were their factors, Mr. James Purvis and Mr. William Darling, both of them Berwickshire men. The red or roan crest of Sittyton now marks the monarch of many a herd; and from cows of shorthorn crosses, Anguses, Galloways, and Shetlanders, which Mr. Dudgeon (now of West Lothian) first introduced at Greenland, the yearlings and two-year-olds spring, of which about seven hundred are shown at Georgemas. These are generally only the second-class beasts, as the best lots are seldom pitched at all, but lifted at once from the farms. Many of the yearlings leave little short of a pound a month behind them in the scrips of the higher feeders. They are taken by the dealers to the Muir of Ord, and find their way from thence chiefly to Morayshire, Easter Ross, and the Black Isle.

But whilst we are wandering off to Georgemas in the spirit, we are really drawing near Barrogill Castle, the home of the Lord-Lieutenant of the county. A rustic cross on a mound attests the friendship of the late Earl of Caithness for his relative George Canning; and the reclaimed wastes and the new steading at Philips Mains prove that the present peer is no laggard in his generation. His lordship is also a great mechanic, and the steam-carriage which he steered so deftly, with the countess by his side, from Inverness to Barrogill, contrives "a double debt to pay," and is now working as a stationary engine at the stone quarry, and yet ready to take to the road once more at less than an hour's notice. Art has here, as in Shapinsey, triumphed over nature in the matter of trees, which formed more than one of those leafy alleys which we had missed so sorely for weeks; but it was not until we reached Barrock, and its snug beltings of thorn, ash, and elm, where the red berries of the rowan-tree are vying with the graceful clusters of the laburnum, and the purple beech with the black Austrian pine, that we began to doubt the saying that there is "no natural wood north of Berriedale." It is given to few men in Caithness to sit under the shadow of the trees they have planted; and this felicity has been attained by Sir John Sinclair (whose zeal for improvement is. nearly as ardent as his great namesake's), through trenching, draining, and enclosing, in the first instance, and then forming plantations in masses, and transplanting from them after eight or nine years' growth. Sheep and cattle both thrive bravely with such shelter; and to this, and a very generous diet, "never allowing the animal to retrograde," no medicated food, and no bleeding and physicking except in cases of unmistakable inflammation, much of Sir John's success as a breeder may be traced.

It is nearly thirty years since he followed in Mr. Horne's track, and introduced shorthorn bulls principally from Thornington and Sittyton. Galloway cows were his fancy a few years later, and from one of them he bred Mr. Owen of Blesinton county Wicklow's gold medal bull. Mr. Miller, of Lower Downreay, still holds by "the heavy blacks," and uses an Angus bull; but Sir John merely keeps black cows, gentle and simple, as the two breeds are sometimes called for distinction, for the sake of a capital cross with the shorthorn. The Galloways are perhaps better suited to a county which has, as a herdsman observed to us, "sometimes three climates a day." Both the first and second crosses with the shorthorn bull lean a good deal to the female in shape, although the black coat is not unfrequently transmuted into red or bronze. They seldom put out horns before the third cross, and then very often mere scurs; but at this stage the shorthorn fairly wins the colour, and the white nose comes as the fringe ear vanishes. Beyond the third cross breeders very seldom venture, as the produce is too delicate and leggy to breed from. Sir John does not keep so many cows as Mr. Henderson, of Westerseat (who has generally 300 acres under turnips on this and his Bilbster property, much of which is reclaimed from moss); and he makes up his numbers with calves from his tenants, who have the use of his bulls. These tenants' cows are a mixture of West Highland, Orkney, and old Caithness with shorthorn; and about a fourth of the yearlings, which Sir John sold last Georgemas at £14 each, were full of Shetland blood on the dam's side. Sir John has had very few Shetland cows through his own hands; but a trio which he once purchased for £7 10s. brought him three calves each, and averaged ten guineas when sold fat. Of late he has principally depended on Sittyton for his bulls; and it was from there that he purchased Malachite, who formed for us, with Whipper-in, Forth, and Royal Butterfly 11th, quite a chain of first-prize Royal English bull-beacons between Barrock and Keir.

Grass in Caithness is generally ready towards the latter end of May, and continues up to the middle of

September. The long gap is very hard to bridge over, and high feeders have sometimes to help out their swedes with strong supplies of corn and cake. No county carries out more consistently the late Mr. Boswell, of Kingcausie's maxim that "you must feed from the very starting-post." It is the general practice of the country to rear calves by hand; but both Sir John and Mr. Henderson consider that a quart from nature's bottle, and taken at the calf's own pleasure, is more valuable than twice the quantity gorged three times a day from a pail. Hence their calves are dropped early, and suckled till the Wick herring fishing begins in July, which gives them from three to four months with their dams. At Westerseat ten or twelve of the best queys are always put to the bull at fifteen months, and are allowed to suckle their own calves the first season. Half of the remaining cows and three-year-old queys come within the milkmaid's province, and the other half have, if possible, two calves put upon them. In winter they are never let out before nine o'clock in the morning, or allowed to touch a turnip with the frost rime on it. The system of Mr. Henderson (whose nephew at Stemster also stands high as an agriculturist and exhibitor of stock) is so good, and so well put in the *Irish Farmers' Gazette*, that we may well quote it in full:

"After being weaned, the calves are turned out on a piece of old grass which is kept for the purpose, but are always housed at night, and vetches or the second cutting of clover given to them, with half a pound each of artificial food, consisting of a mixture of bruised oats and oilcake. This quantity of cake and corn is continued until January, turnips and straw having meanwhile taken the place of the green food, when the allowance of mixed oilcake and oats is increased to one pound per head. In April it is again increased to two pounds each, with hay or oat straw, until the grass becomes ready for cutting. It must be observed that the young stock do not get, during winter, as many turnips as they would eat. The first feed in the morning is their allowance of oilcake and oats. At 9 o'clock a.m., turnips are given, and again at 3 p.m.; but there is no restriction put on the quantity of straw, and what they do not consume between the feeding times is used for litter. In addition to keeping the young stock in a constantly improving state, which is a most important consideration in any system of rearing, the Westerscat mode of feeding has proved to be a complete antidote to the quarter-ill or blackleg, which at one time prevailed on the farm. Calves which have been kept in the manner described never lose condition, but are always getting better, and the result is that they fetch from £12 to £15 each when a year or fourteen months old. A lot of twenty-four of that age were sold last April off the farm at £13 a head, and were considered cheap at the money. If kept on till they are eighteen months old they will weigh 40 to 48 stones—that is 5 to 6 cwt.; and such have been sold by Mr. Henderson in October at £18 a head. The usual quantity of turnips given to young beasts is about 40lbs. per head daily. Some keep their young stock tied up in the house, turning them out for an hour or two into the foldyards; while others keep them in open courts or yards, having sheds attached, to which the cattle have at all times free access. These yards hold five or six beasts, and great care is taken that the partners in each court shall be equally matched; for it would never answer to have a weakly animal put among others which were stronger."

Sir John's system differs from Mr. Henderson's in this, that the calves are weaned on 1lb. of oilcake, and as soon as the weather is cold and they come into the yards the allowance is increased by 1lb. of bruised oats. Where the calves do not suckle, each has about eight quarts daily of warm milk direct from the cow, and divided into three meals. After the first three weeks, a little oilcake made into jelly with hot water is put into the milk, and gradually increased in quantity up to a pound per day. When they are nearly four months old, skim-milk is substituted for new, and they are carried on with it and the oilcake for a month, and are then well grazed and generally sheltered at night.

Sir George Dunbar was, like Sir John Sinclair, one of the first followers of Mr. Horne in the shorthorn crosses; and we found some promising prize pairs in his meadows at Ackergill Tower. During the earlier times of the local agricultural shows, he carried off the prizes for a series of years with one and two-year-old crosses. He has since then been successful with the cross between Shetland queys and shorthorn bulls, selling the bullocks for 11 gs. at sixteen and eighteen months old, and keeping the quey calves to try a second cross. Latterly Sir George has given his attention much more to breeding Leicesters. He began with ewes purchased upwards of thirty years since by his father, through the English "Nestor of Shorthorns," Mr. Wetherell, from Mr. Davidson, of Cantray. To these succeeded some ewes and a tup from Mr. Compton of Learmouth, and then a lot each from Mr. Thompson, of Haymount, and Mr. Cockburn, of Sisterpath. One of the Brandsby tups did not answer, as Caithness declared itself early in the day quite as strongly as " the little kingdom of Scotland and Northumberland" against the blue-faces.

The size of the Border tups was what the breeders yearned after as well; and hence Sir George has generally gone on the system of buying a couple every year in the Kelso ring. Lord Polwarth's round-ribbed and flat-backed type is what he has held by, and he was the last bidder but' one for the top £62 shearling of " 'Sixty-three." The clay loam of Caithness suits turnips, and wool-staplers aver that the clayslate rock, which in many parts lies close to the surface and sorely foils the drainer, communicates a peculiarly lustrous quality to the fleeces. No

E

wool commands a higher price than the Caithness in the auction-room, and Sir George's gained a gold medal at the Great Exhibition of '62. His pure Leicester flock numbers about eight score, and have the run of the fine pasture land round the Tower. The waves beat up to the base of this massive keep, which on one side breaks the outline of Sinclair's Bay, and on the other looks out on choice vineries, and green-houses, rich ribbon borders, and two ancient dovecotes with acacias twining round them, and prize rams busy with "their little white ivories" below. About fifty shearlings were sold last year at five guineas each, and several of them went into Inverness-shire and Aberdeenshire. In '63 a lot of twenty departed for Edinburgh, and fetched the second highest average at its first tup sale. With the exception of Mr. Brown, of Watten, there is no other Leicester breeder in the county, and between them and a few arrivals from the Edinburgh and Kelso ram sales the ewes of Caithness find partners enow.

The mail ride of the year before, from Wick to Golspie, had been novel in its way. In this era of "Stokers and Pokers," and Post-office vans pierced inside with a hundred labelled holes and armed outside with a cunning catch-net, it was a relief to renew one's youth, and sit behind with the guard. He pitched out his bags with admirable precision; he caught up others from the point of the official-forked stick; he dispensed his nods and wreathed smiles to " nearly seventy miles of females," until it grew dark and his fascinations were lost upon them; and he treated us in the intervals of business to divers essays, clerical and lay. No man suffers less, apparently, from leakage of memory. He knew something about every one, from the minister who came and met his friend, to the bride of three weeks on the "knife-board." One of his primest August uses seemed to be carrying news of the overnight's fishing from village to village; and an emissary of the beach was always starting up at some corner of the road with the eternal cran query. Fishing-boats were boxing up towards Wick, either to verify the news of a good bank of herrings or to try and mend their luck with the Lammas Stream. A yacht raced us, and we never could shake her off till she took to tacking near Helmsdale, and then stood across the Firth. The road was dreary and Irish-looking enough for some miles out of Wick. A bourtree or elder-bush attached to the tumble-down cottage compound was the only shade till we reached Lybster, where the nightingales with which the late Sir John Sinclair once vainly strove to make Caithness vocal might have trilled their notes in a pleasant grove. A yellow caravan was bearing along a dwarf and "the Hottentot Venus"; little girls girded their kirtles up to their knee, and chased us for fully two miles, as clear-winded as "Deerfoot" to the finish; but the poor idiot, who watches year after year at the same gate for the up and down mails, had grown so plethoric on coppers that he could hardly stay two hundred yards for all the guard's spiriting. Three shepherds and four dogs came in procession to see a pup off in a basket; and of all horsekeepers we voted David Ross, of Begg, quite the sprucest and the best. Well might we contrast his style of turning out his team, and his anxious consultation with the coachman about a stable invalid, with that of the boor who had left the shoeing pretty much to take care of itself, and who blurted out, when he was begged to be "canny" with the leader, "*It's no the flies—it's bad manners.*"

On the occasion of our second visit, there was nothing for it but to steel ourselves against all bed regrets, and face the muirs at night. We wended our way by a series of zig-zags across from Barrock to the high-road between Thurso and Lybster, and then struck straight for the coast. The sun went down, and the rain took no half-measures with us. Soon every door was barred, and every light put out in the few cabins along the road; but one family at last responded to our hail with biscuit, cheese, and milk, besides offering a bed and abundance of tares. There are many dreary passages in a man's life; but wiping down a mare very short of condition in your shirt sleeves and a cow-house, on a wild muir, by a dim, spluttering dip, at midnight, with the wind sighing through the broken panes, the heavy rain-drops pattering on the door-sill, and a forty miles' ride before you, has very few to match it. Still it had to be done; and "if I mun doy, I mun doy."

The mountain burns, which soon began to run right viciously, made "music in our sorrow"; and as the moon sailed out from behind a cloud, and shone on the long pools which were fast gathering by the roadside, they seemed like polar bears craftily stealing along. We hailed the mail-road at Lybster and the roar of the sea as quite old friends, and felt a little comforted. As for the mare, she was like a whole troop of them rolled into one. Though she had only been with us two days, she had got so accustomed to our voice, that, if we fell a little behind, she would stop when she was spoken to, and look round, first to the near and then to the off-side, in the gloom, to be sure we were at hand. Weariness at length defied all face-washing at the roadside springs, and two hours of that night are best accounted for in the preface. Be that as it may, the mist wreaths began to curl lazily up the deep mountain ravines, and away to that vast, granite deer "forest" behind. Morning broke, and the rain was gone, and the rainbow was spanning the Berriedale valley. There were all the varied purples of the heather, and the rich green livery of fir and larch, to brace us up for the dreaded Ord of Caithness; and the mail, as it rattled cheerfully past us, was quite "the missing link" with mankind. Morayshire, on the opposite coast, looked like the outline of a new world, beyond a calm, blue-dimpled sea; and as we rounded the last crag near Helmsdale, the gently curving sands of Sutherland lay at our feet, white and warm in the early sunshine.

CHAPTER IV.
HELMSDALE TO MEIKLE FERRY.

"He once asked a roomful of divines why white sheep eat so much more than black sheep. One person advanced it as his opinion that, black being a warmer colour than white, and one which never fails forcibly to attract the sun, black sheep could do with less nutriment than their white contemporaries. At all these profound speculations Dr. Whateley shook his head solemnly, and then proceeded with imperturbable gravity to explain—"White sheep eat more, because there are more of them."—ARCHBISHOP WHATELEY'S LIFE.

A Walk down Kildonan—The Cheviot Flocks of Old—Foxes on the Gibbet—Mr. Hadwin's Stags' Heads—Gideon Rutherfurd—Kildonan Churchyard—Sutherlandshire Flockmasters—Shows in the County—Sheep Farming—Heather Burning—Sporting in Sutherland—Strath Brora—The Dunrobin Herd—The Museum and Kennels—The Dunrobin Dairy—Climbing Ben Vraggie—The Meikle Ferry.

HELMSDALE, which was built by the late Patrick Sellar, is not exactly a paradise amid the moil and offal of the herring season, and never was change more welcome than a walk down Kildonan. Seagulls seemed to be acting as rural police over acres of nets which were laid out to dry; while a black bull from Dunrobin, with that especially royal sit of the head which marks the West Highlander, stood and looked the very warden of the strath. We followed the course of the Helmsdale river, where an angler (dressed in a suit in which his college dons would have disowned him) was patiently guiding his fly; and found the ground all lea, with a strong tendency to fog and rushes. Some of it gets top-dressing, but it is terribly rough, and the Norwegian harrows will hardly touch it. The roadside was almost honey-combed with rabbits, which lead far too merry a life of it among the fern. In fact, we hardly heard a shot all day, and were too late for some capital wild-goose shooting on a chain of lochs farther up. Thirty brace out of some five hundred fell to six guns; but they are very hard to hit, and when they are disturbed they will dive and float along with their dark nebs just above water, and if they do take to the heather they run as low as an otter.

Once upon a time, nearly the whole strath was held by the late—Mr. Reed and Mr. Houstoun. The former, who had 18,000 sheep, began at Edruble, about four miles beyond Helmsdale; and with the exception of Major Gilchrist's farm, he occupied the whole eighteen miles by eight on the left bank of the Helmsdale up to Aulton Down, the point where it joins the Brora. His farm also ran with the river down Strath Brora; and he held Balnakiel in Durness parish, near Cape Wrath, as well. With such scope and choice of pasturage, it is no wonder that for nineteen successive summers the celebrated Jamie and Watty Scott of Hawick wrote and took about 2,006 wedders and 1,500 cast ewes. The late Duke of Sutherland gradually broke up this gigantic flock system. At first the rents ranged from eighteen-pence to half-a-crown a sheep; but, since then, the rate has increased on many farms to five shillings, and three only obtains in the current leases. The strath of Kildonan springs early, and there is nothing in the lower part under three shillings, and nothing in the higher under five.

We did not pursue our own wanderings beyond Mr. Hadwin's shooting-lodge, in front of which two foxes, killed in no fair chace, and then stuffed with straw, creaked like felons on a gibbet. No Oulton Lowe or Cream Gorse for them; no cheery view-halloo from Jem Hills or John Walker! But there was no help for it; the cubs cannot be dug out of the rocks, and the old ones make as much havoc with the lambs and hares as they did of late among the roe-calves in a Morayshire wood, and "*grew too fat to trot.*" Still extremes will meet even in appetite; and if they do love one trap-bait better than another, it is a rotten rat. There were far more pleasant trophies inside the lodge, where scores of antlered heads did peg duty in passage, hall, and bedroom. Many of them had their forest epitaph duly labelled below. August, 1859, had witnessed the crash of a 19st. 2lb. hart on Ben Dunc at 295 yards; Glasscome could boast of "24 stone clean"; and Aultongrange beat it with a stone in hand.

Old Gideon Rutherford, whose son Richard now holds part of it, is the venerable sage of the strath; and there was the light of battle in his eye, when he recalled how many "braw lads" it had "given to the 42nd Highlanders, who beat the Invincibles in Egypt." There have always been men of sturdy fibre up Kildonan. One of them, when past eighty, was begged to keep warm in his cabin during his last sickness; and yet he would argue, "*It's keeping fra the wet that maks me ill.*" "The Sheeperd Swingle over the remains of his beloved son" raises a simple stone; and his own name is not found on it till he is 107. Nettles and foxgloves were growing over the threshold of that deserted church. The door was off its hinges, and the hole in the floor of the pulpit told how the portly minister used to emphasise Sunday after Sunday with his right foot. Earth lays in turn its heavy load on him, and he sleeps among his shepherd worshippers with "Mistress Isabel and Jean, late deceased spouses, both judicious, worthy, and amiable characters, and attentive to the interests and comfort of himself and his children."

The flocks are larger on the west coast than the east, and some of them muster nine thousand or more. No patriarchal pride seems to be felt respecting their extent; and the owners are wonderfully taciturn and costive on the point, though the salesmen form a pretty accurate notion from the annual ewe drafts. We should as soon presume to reveal these crook mysteries as to report the general wedder criticisms of the Cumberland and Yorkshire feeders—how some are becoming rather fine-bred; how others are grand in size, but not such good travellers; how a third are good grazers, but going off in size, and so on through all the Cheviot chronicle. Still the critics seem to hold the balance very equitably between the Highland and Lowland flocks. On the west coast of Sutherland they are all wedder farmers; and Patrick Sellar of Strathnaver (who has now the farm in Lewes, which Wattie Scott held for nineteen years), Patterson of Mellness and Bighouse, Clarke of Eribol, Gunn of Glendhu, Clarke of Stronchrabie, John Scobie of Lochinver, M'Kay of Kinlochbervie, Reed of Balnakeil, and Sangster of Coignafern make up a strong bede-roll. On the east coast, the ewe and wedder farmers are nearly equal in strength. John Hall of Seibers Cross, Reed of that pleasant spot down Strath Brora, which the gods call Kilcalmkill and men Gordon Bush, Houstoun of Kintradwell, Marcus Gunn of Kilgower, Murray of Kirkton (whose ewe draft goes annually to the Royal Home Farm), Andrew Hall of Blairich, and Hadwin of Kildonan keep wedders; while Dudgeon of Crakaig, Hill of Navidale, Rutherford of Kildonan, George Ross of Torboll, Barclay of Davochbeg, and Major Weston of Morvich have ewe farms, and sell their wedder lambs.

The East Sutherland Farmers' Club holds its show at Golspie for sheep, West Highland cattle, ponies, and pigs; and the men of the West decide their wager of battle at Aultnaharrow. The combined county show is at Lairg, where a silver medal is given by the Highland Society, open to proprietors, factors, and farmers, for tups which have won a prize at a previous competition. The Sutherland men care very little for showing out of their county; and at the Highland Society's annual meetings, the Lowlanders—with Aitchison, and then Brydon, as their champion—have had the Cheviot classes pretty nearly to themselves.

The late Mr. Reed, of Gordon Bush, was the first Sassenach who brought the Cheviot into Sutherland. His flock came with him from Reed Water, on the South side of the Cheviot Hills; and when the Robsons followed him, the knell of the black-faces in Sutherlandshire was rung, and there was no farming against the Borderers. Although in the shepherd's mind the Sutherland sheep live on heather, and the Border ones on grass, the latter changed the venue in one respect for the better. The "cotton plant" or mossy grasses in the lower range of Sutherland lie very little above the sea level, and tide the sheep through the winter and spring months, when those on the Border hills are generally hid in snow-wreaths on the summits. This plant is, in fact, as much the making of Sutherland as its prototype is of Manchester. On the west coast more especially, deerhair follows the mosses most opportunely for six weeks before May-day; and the "flying bent," sometimes to the extent of hundreds of acres, is won and built each August into piles on the muirs, to feed the sheep with in winter. Wedders take the stormier ranges; and in a very severe time, the sharp spirals on which they mainly subsist bring on pining, and nothing but a change of food restores them. The black winter of 1859-60, like that of 1838-39, was very equitable in its ravages. Some Sutherland farmers lost nearly a third of their flock; and the ewes got so low, for lack of food on the hill, that they were mere cabers in point of milk, if they did bring their lambs. They had hay for more than eight weeks, and in many instances sheaves were let down to them, and scores of the most weakly just struggled through on warm oats and hay. In England, the hoggs do not rise to that rank till they have been clipped; whereas lamb promotion is speedier in Scotland, and takes place about October 20th with the smearing. Olive and castor oil have been tried instead of the conventional butter and tar; but although the wool brought a higher price, it of course weighed less, and rather impoverished the mutton by not turning the rains of winter and spring.

To prevent braxy, which is generally induced by eating diseased vegetation during the frost, the wedder hoggs are turniped, in or out of the county, from the beginning of December to the end of March, along with the tups and the worst of the dinmonts, to make them equal to the lot on the hill. The wedders are always sold (like the ewes, of which about a third are cast each year) by the clad score, or 21 as 20, and if they kill to 20lbs. a quarter, as three shears, after four months on turnips, it is considered capital. The bargain is generally made at the Inverness Character Fair, where the Cumberland, Dumfriesshire, Yorkshire, and Lothian men buy most freely; and several lots change hands at the September Falkirk, on their way south. The closing chapter of wedder history comprises another course of turnips for three or four months, and then many of them are consigned to the Liverpool and Newcastle fat markets. When the brothers Scott were in the height of their trade, and the markets happened to be cheap, they would buy more than forty thousand wedders and cast ewes on the plane stanes at Inverness, in addition to those of their own breeding, and pitch the majority of them at the September and October Falkirk trysts.

A barren ewe is marked with ruddle on the back of her head, and the token is renewed with tar at clipping time, which enables the shepherd to put her among the draft, if she misses the next year. The trace of the black-face sometimes peeps out in horns or a black foot and ears; but, provided the bone is nice and sclef, the breeders do not dislike grey legs, and consider that they indicate good provers. Red or white noses or a pink ear invariably show softness, and for this and hairy wool there is no benefit of clergy. The weight of the hogg fleeces depends so entirely on their keep that it is difficult to strike an average, but even fine sheltered turnip land cannot send them to scale above $4\frac{1}{2}$ to 5lbs. A three-shear ewe which has brought up a lamb will average 4lbs., wedders of that age 5 to 6lbs., and tups which have been on turnips all winter 7lbs. The wool varies with the winter, and in very severe months it hardly grows at all.

Shepherds begin as lads of eighteen, and serve a four or five years' apprenticeship, at £16 to £25 wages and their keep. At the end of that time, if they become master-shepherds, they have a cottage and grass for two cows and a horse, and a pack of eighty sheep. Sometimes they have only half-a-pack, and the other moiety in wages. The pack is quite a miniature flock, with a separate mark, and the cast ewes go with the master's. They also get an allowance of 6½ bolls* of meal for themselves, and the same for their lad where it is "a double herding," to wit, 1,000 sheep, "be the same more or less." Their two-acre farm is generally in a course of cropping for potatoes or barley, and a few turnips for the broth-pot; and a braxy victim, when it has been skinned, well pressed with stones in a burn to extract the inflammation, and then salted, makes no contemptible hung mutton. The shepherd has a hard six weeks of it, to see that the ewes do not wander away from the tup; but it is nothing to the lambing season, which begins about April 18th and lasts for a month. Fifty ewes are generally apportioned to each tup on the hill, who needs to be a good traveller, and is generally preferred when two years old. No one save Mr. Dudgeon, on his fine seaside farm of Crakaig, tups his gimmers as

* A boll = 140lbs.

a rule; but some will select a score or so of the strongest, and put their lambs on ewes which have lost their own. The ewes are brought on to the strath to lamb, and in fair average seasons eighty-four lambs to a hundred ewes is not bad. There are different bits in the marking system of the east and west coasts; and on the latter, where the flocks are more liable to be mixed, the face is marked as well, at an agreed angle. If, on the east coast for instance, a man's land marches with another, he will mark his ewe lambs with a double forebit on the near ear, to begin with. The next summer it is double backbit on the same ear, then ditto on the off ear, and so round to double forebit again. The wedder lambs are marked in this ear order, but on the single-bit system, and his neighbour pursues the same process, but exactly reverses it.

The conflict between the sheep and the shooting interests is sometimes pretty hot over the heather burning, as the shepherd is anxious to burn as much and the shooter as little as possible. The old system was a peculiarly expensive one, as a whole cohort of broom-men had to be paid and kept for several days, waiting for the wind to set in from the right quarter. Mr. Houstoun's system, which has been adopted to a slight extent by Colonel Hunt in Kildonan, is far more economical and effective. The heather is cut into squares by means of drains 3 inches deep and 28 inches broad, and burnt on a three or a five course rotation. Thus the reversed sod goes to make up a

56-inch barrier, and prevents the fire from spreading down wind, while the drain furnishes drink to the grouse in a dry season, and is never too deep to drown the broods. The squares in all their varied rotation hues look very beautiful on the side of a hill, and in the third summer after burning they will always hide a grouse.

The coast road from Helmsdale to the Mound, with its dark masses of crag and verdure, is quite our favourite stretch upon Scottish ground. We like to scramble about the hillocks at Crakaig, and watch the Shetland cows, with their lusty, long-haired, black-roan calves, feeding in a troop along the beach, and a shorthorn yearling bull in command, as white as Europa's love. The evergreen gorse, from which the Kintradwell Clumber scorns to flinch, flourishes with the wild willow hard by those rabbit-haunted links. The otter glides up Collieburn, and calls after its loch-fishing, to enjoy a trout under the ledge of the bramble-fringed rocks, where the kestrel has its eyrie, and from which the wood-pigeon sweeps forth with a strong rushing flight, as if an East Lothian price were set on its eggs and head. Partridges scurry along among the sand-hills, where the sheldrake builds in the deserted rabbit-hole, and hatches her brood of sixteen; and the notes of the plover, the oyster-picker, and the ring-dotterel break in all their varied cadence on a naturalist's ear.

As night draws on, sea-birds of every form are "crushing the air to sweetness" with their strange fishing-cries; and the cormorants choose their stones as gravely as if they were Martello towers, and themselves part and parcel of a regular coast-guard. Even the mild porpoise of the beautiful wave-line is busy chasing the herring-fry; and the seal, which is whelped with a brain as big as Bunsen's, or "plain Jock Campbell's," in some rugged Caithness or Iceland cave, glides quick as a shadow through the waters to the Brora mouth for a salmon, and will hardly quit it for small shot. The salmon spawn from the bridge of Brora, and when the water is warm they take the upper fall to Benarmin, at the foot of whose deer-forest the north branch of the Brora comes out of the springs. Strath Brora has not the width nor the same amount of grass as Kildonan, but it has far more for the eye in its richly-tangled copses of mountain birch, with their ground-work of faded breckan. The stillness of its loch is unbroken, save by the jump of a trout or the ever-widening ripple of the golden-eyed duck. The Carrol Rock stands out boldly like a bastion above the Duchess's drive, dwarfing the hut of the watcher into a mere speck at its foot, and throwing its green and purple shadows over the waters; while behind it, stretching away in the grim, grey distance to the head of Rogart, is the mighty Ben Horn, that very unicorn of deer-forests. The grave of Malcolm near Kilcalmkill is marked by a few flagstones, and beyond, over ford and fell, is the lonely Seibers Cross, so great in wedder history.

But we retrace our path once more, and bidding a

reluctant good-bye to Kintradwell, we ride on past the deep, copper pool at Brora Bridge, seven miles by the coast to Golspie. Along the hill-side to the right are the cottages with their plots of ground, which were allotted to those Highlanders who would not emigrate when they were ordered to quit the glens. It was doubtless a sharp sermon, and rendered doubly so by the stern opposition; but even the traditions of "hame" will, as years go on, melt before the conviction that chronic snuffing and shin-toasting and rearing a few potatoes within a tumble-down wall are not the mission of a Highlander. They were taken from that useless existence to a spot where they have full exercise for their energies both by sea and land. It was a readjustment, very bitter to the Highland heart, but still wholesome and right, as sheep were placed where there ought to be sheep, and men where there ought to be men.

Beyond Brora we skirt the Uppat woods, and for the first time in our life we scan a roebuck, "a perfect form in perfect rest," actually standing motionless under a fir-tree at twenty paces. With all the wisdom of the rook, it calmly surveys our gunless friend, William Houstoun, who, almost tortured to frenzy at the sight, gasps out that it has a three-year-old head, and begs us, out of sheer mercy to him, to halloo it away. West Highlanders, black faces, and "ponies for the hill" are the joint-tenants of the Dunrobin policy; and a turn up an avenue to the right brings us to the castle steading, whose dun and brindled breed has long been known in the land, and fetched top prices at the northern markets for nearly two hundred years. The herd consists of thirty-four pure West Highland and twenty-three cross-bred cows. Some of the latter are between the West Highland and the Ayrshire, which makes a rare first cross, taking rather to the Ayrshire, and gaining in quality of milk what it loses in quantity. Comparatively few Ayrshires are kept in Sutherland, as the climate is rather too keen for them, and both for flesh and milk the first cross between the West Highland and the shorthorn is the favourite. The Dunrobin system is to let the calf have one side and the dairymaid the other; but if the West Highland cow is a crack, she and her calf wander off to the Big Burn field, which gives them a rough bite all the winter; while the cross cows are kept in the house, and their calves reared by the pail. The former are brought in for a day or two before calving, to accustom the calf to be handled; but that is all the shelter they get. In coarse weather the bullocks have an open fold, and, with turnips and a little cake to help, they are generally killed when rising four, at about 11¼ cwt., with "a good deal of the steelyard inside." The first-prize cow at Paris in '56 was bred by Mr. Stuart, of Duntulm, and prepared for show by Mr. Tait, who became Her Majesty's Home Farm bailiff about six years since. A dozen beasts were added from the Breadalbane sale; and the brindled ox, which was third to the late Duke of Athole's yellow and dun

Breadalbanes at Kelso, was bred by Dr. M'Gillivray at Barra. The herd are not intended for showing, but to keep up the breed or a stout cross; and five or six bull-calves are reserved every year to fill up the ranks of the thirty, which are dispersed on service from July to October among the tenants, as far as Tongue and Lochinver.

The head of one of them, with horns three feet from tip to tip, guards the entrance of the Dunrobin Museum, which is quite a key to the natural history of the country. Assynt seems to have been a fatal haunt both to the fox and the ring-tailed eagle, which is irreverently labelled "carrion." Near them is the wild-cat, one of the old die-hards which haunt the cairns in the snow time and form the Duke of Sutherland's crest. The marten-cat is nearly extinct; but a polecat specimen is more easy to get, when it has been routed by the aid of a muzzled ferret out of its winter magazine of broken-spined frogs. Jacko, the monkey, and late of the duke's yacht, is the comic countryman of this still-life piece, and, with upraised eye-glass, he cons "my marriage lines." Feathered wanderers from over the sea—the roller (which looks like the kingfisher of the east), the hoopoe, and the Bohemian waxwing—have all come to grief on their arrival, and share this dainty Morgue with every Sutherland bird; and crystals from Dunrobin Forest mingle harmoniously with an elfin arrow-head of flint, and a ball from Montrose's last battle. The very wasps are represented by their filagree home; and each bird's nest in the county must have been rifled once on a time, to furnish materials for egg lore. They are to be found of every hue and size in those cases— the large balloon-shaped chocolate with black spots of the guillemot, the white-marble ones of the wild-goose, the round and pale-blue hieroglyphics of the curlew, the snow-white ellipses of the rock-pigeon, and so gradually down in the scale to the white with brown dots of the willow and the still tinier products of the golden-crested wren.

Forp, the dog of Sutherland, which is supposed to have run a match with the best of Ossian's, has a stalwart namesake in the kennels (whose yards are built on the slope); but though his stern is that of a half-bred, Macdonald the keeper assures us that he " can make a deer's bones crack again." Hector, on the same authority, " always takes them behind"; and Fan, who has more of the bloodhound in her features, has lived, like him, to prove that it was a good day's work when their puppy lives were begged at Meikle Ferry, and they arrived at Dunrobin in the pockets of Macdonald's velveteen. The rifles, German, Purdey, and Double Lancaster, which stand in such tempting array, along with the deer-saddle and all the other belongings of hill craft in the gun-house, have not for ten seasons past given an account of more than "23st. 2lbs. clean." This deer was shot at Balblair Wood by Lord Delamere, and is only *proximus intervallo* to those of 29st. and 30st. clean, which are credited at Dunrobin to Mr. Holford and the late Lord Ellesmere. The venison house, which resembles a small chapel at first sight, is a most happy combination of thorough draughts, Welsh slate, and polished pine, and has space for five deer and sixty-two quarters on its pulleys and side-hooks.

The Duchess's dairy is across the road, in the glade beneath the castle. The bunch of heather in the grate, and Landseer's Milkmaid over the chimney-piece of the sitting-room, are in quiet keeping with the white delf-bowls, the butter-pats floating leisurely round the water-lily fountain, and the rotatory oak-churn with its burnished brass hoops. A flight of rustic steps, thick with honeysuckle, leads to a balcony in the steading. There are stalls below for twelve half-Ayrshires, with what look like wine-bins at one side for their calves, and a still more suggestive milk-hoist for the visitors. Ribbon-borders with the pale pink saponarium, the white nemophylia, the blue salvia, gardener's garter, and the never-failing variegated mint, run coyly from the castle gardens to the sea. Fuchsias, hollyhocks, and dwarf dahlias blend with "the red, red rose" and the hardy spray-sprinkled green of the buckthorn; and we hail the presence among them of M'Alister the duke's piper, not for his family skill, which falls dead on English ears, but because he is the only Highlander, save two, that we have met in full costume north of Inverness.

Then we are seized with a desire to scale Ben Vraggie, and find that reaching the Duke's Statue by a short cut through the deep dingles of the Mound Wood, so dear to woodcocks in the season, is certainly not the shortest way; but the view of the Moray Firth, and the whole of that silent coast along which we have wandered three summers, and felt each time more loath to leave it, soon made us forget the toil and the brambles. It is easy enough sliding down again on to the clay lands of Kirkton, and then comes a very Mamelon of a crag, beneath which Hugh Miller, in his mason days, is said to have first pondered over the Story of the Rocks, within sight of "his own loved Cromarty."

The goats and kids can hardly crop the lichens and moss on it, and, falling over, leave their skins to their country as purses for Highland regiments. But the sun is sinking behind a chain of hills, miles away near Strathfleet, and Tain is the goal to-night. Quiet little Dornoch, with its cathedral, the resting-place of the old Earls of Sutherland, is hardly visible on the left, and the dreaded Meikle Ferry is in front. We have seen it under three different aspects: once when the rowers had "been i' the sun," and the coachman, guard, and passengers were all fain to take an oar; again, when Croall and Son had set up a small steamer; and on this occasion, when at the fifth and last attempt our recusant mare was

(N)

bundled into the boat like a sack. No wonder the process had haunted us, by anticipation, all day; but once in, we had nothing to do beyond deadening her ears against the sail-rustle, and she "*jumped out sharp*" into Ross-shire at last.

CHAPTER V.
TAIN TO INVERNESS.

> " And then along the plane stanes
> Like a provost he would go;
> Oh! we ne'er shall see the like
> Of Captain Paton no mo'."
> PROFESSOR WILSON.

The Black Dog of Tain—A Jockey Club Wanderer on Morich Mhor—Ross-shire Tod-hunters—The Crofters—Flocks in Easter Ross—From Tain to Dingwall—Coul Cottage—The Black Isle—Belmaduthy—The Muir of Ord—The Caledonian Boots—Highland Society's Shows—Inverness Character Fair—Hugh Snowie's—Stags' Heads—Laggan Cottage—Nairnshire—Lord Cawdor's Old Scotch Flock—Kildrummie and Lochdhu—The late Hillhead Herd—The Witches' Heath.

UNSETTLED spirits have been said to roam the earth in the shape of a black dog. At all events, we were formally taken in charge by the latter when we reached Tain; and it never left our side till it lured us, as it had done scores of visitors before, into its fancy baker's, and made us buy it a bun. It was the only black mail levied on us throughout our wanderings, and as such it deserves recording. Before the railway days, it was easy enough to find your way into Tain from the ferry by moonlight, but even in daylight it is not so easy to work outwards. What seems the straight road to Inverness really leads on to the Morich Mhor, a large sandy flat, where the

Duke of Gordon, Lord Lovat, Mr. Davidson of Tulloch, and others held races nearly fifty years ago. This illusion, or some unconscious sympathy with the spot must have led the late Mr. Charles Greville to stray three miles down it, on his return from Dunrobin, while the mail with his valet and portmanteau was stealing ahead through the pine woods of Calrossie. Luckily Mr. Kenneth Murray was driving home, and, seeing an old gentleman quite beat with fatigue, offered to take him back to Tain, little knowing till they exchanged cards what an eminent pilgrim he had succoured. The Mhor, as you see it from the high road, strongly resembles the ground of the Waterloo Cup Wednesday; and it has been tenanted by many a dark-coloured, straight-backed hare, as Gordon Castle greyhounds have known to their cost. The Calrossie Woods produce rather more mixed sporting, and about one thousand squirrels and two hundred and fifty brace of woodcocks were killed out of them recently in one year alone.

The cultivated part of Ross-shire is a mere fringe of the Moors, and the other nine-tenths of the county up to the shores of the Atlantic are devoted to grouse, sheep, deer, and crofters. The last-named have few sources of existence save fishing. If they are seized with "the low ambition of rising in the world," and come East for the purpose, they make good workmen, and are content like the rest of the labourers with their cup of coffee; but their Gaelic tongue is sadly against them. The tod-hunters, with their terriers and their decoy-fox, have plenty to do in Ross-shire, and sometimes extend their labours to Lewes. These "sheep farm police," as they are termed, would have to become a great band, and watch men as well as foxes, to be thoroughly worthy of their title; as a large ex-sheep farmer, who can now look back and smile at his losses, states deliberately that he never could account for some hundreds annually out of a flock of eight thousand—except in one way. The crofters on the east coast of Ross-shire, or "Easter Ross," are a superior race of people, who have picked up the Scottish tongue, and hobble as well as they can after the times. Like the farm servants, they generally keep a pig, which journeys with a rope round its waist to the local market at Kildary, when it is nearly a year old, and is bought by jobbers for the south. Here, as, in fact, throughout nearly the whole of Scotland, there is among the poorer classes, not exactly a Judæan detestation, but an aversion to ham and bacon. They will rear it, but only for sale; and the home commissariat is more dependent on a bit of mutton, which is either a cade lamb purchased out of the cartloads of Cheviot or black-faced "shots" which the farmers send to the Muir, or a keery bred by themselves, with crosses innumerable. These pets seem to wander over their little unfenced crops just as they like, and to eat out of the pot as well. Their wool supplies the gude wife's spinning-wheel all winter; and two or three of them, a pair of Highland ponies, a stirk or two, and a dozen hens verging on the Dorking, which is supreme in these parts, make up the stock-in-trade of the conventional five-acre holding.

The cultivated parts of Wester Ross have much more pasture land, strong and light intermixed, than Easter Ross, which rejoices in far richer wheat and turnip soils. It is very handy for the glens, where the wedder hoggs return in April, and perhaps supplies more turnips for them than any other district in the North. Its beech hedges and pine woods give it a more cozy look, but still the heavy green and white crops and the engine-house chimney towering above the steadings very much recall the high farming and well-to-doism of East Lothian. Caithness beasts are the farmers' delight. They generally buy the Georgemas yearling stots and queys at The Muir, where their fine scale contrasts strangely with their future companions, the Ross-shire second-class West Highlanders, which are bred by small tenants on the mountains. The light-land farmer carries them on till they are two-year-olds, and then they are finished off on the strong clays. There are no dairy farms in Ross-shire, as the farmers prefer suckling their calves on the Caithness system, and consider early beef maturity a higher aim than butter, cream, and cheese. The Hillhead herd gave shorthorns a great fillip, but the pure sires have to meet a very varied class of dams. Ayrshires are few; Devons and Herefords have hardly been seen in the county;

polls are as rare as a black sheep in Sutherlandshire; and all cattle stock, whether bought or bred, is generally fed off at its third Christmas. The horses are of every variety, but there is not a blood sire within many miles to our knowledge. An active cart-horse called Balmoral left a strong grey foal mark, but the Clydesdales are not great favourites, on account of their "perfectly boundless appetites." Even a Suffolk Punch took the lead for some seasons, and furnished no confirmation of the assertion that this breed is apt to go blind in a northern climate; but he eventually went mad instead, and was shot through a hole in the shed roof. There are scarcely any black-faced sheep in the arable parts of the county, as they are troublesome and get entangled by their horns in the turnip nets. Very few flockmasters keep them, and the general mode of sheep-farming on the hills is a Cheviot flock, from which the wedder hoggs are brought down the first winter for Aberdeen green tops. Many farmers also keep their own cast ewes on the low ground, and cross them with Border Leicesters, for which they give good prices through their agents at Kelso.

The harvest was glowing like a furnace in the rich plain skirted by the hill country, between Tain and Dingwall. Now, we would pass a little tavern where the landlady was scolding a fou customer in Gaelic, and then relaxed for our benefit into the Scottish tongue. A four-penny photographer, with his blue sheet nailed against a cottage wall, the table, chair, and flower-vase all correct, and several women and children as audience, asked touchingly for our custom, and demonstrated how beautifully he could focus us at any distance, by backing recklessly with his camera into a field of oats. There was hardly a ripple in the harbour of refuge, and the hot haze which was driving both dogs and grouse to the water springs almost hid the Souters of Cromarty which guard the entrance of the Firth. Then leaving the fish triangles at the cottage-doors for a season, we strike rather more inland, and the drowsy hum of the reaping machines mark the farm of Teaninich, on which Mr. Tew, the Shorthorn M'Combie of Ross-shire, breeds and feeds his heavy weight champions. Florist came here for a season from Major Wardlaw, but he left very few calves behind him, and was succeeded by Goldseeker (19866), bred by Mr. Wilson, of Brawith. The roan Negotiator, who strains back to Mr. Tanqueray's blood, was the sire of Mr. Tew's best bullocks, which are principally out of shorthorns purchased from Mr. Mitchell of Howgill Castle, in Westmoreland. To make £1 a month from the birth to the block is his aim; and for his thirty months' pair of prize bullocks in '62 he achieved it exactly without any very extra keep. His two pairs of yearling bullocks, which beat Ross-shire the following year, were sent to Inverness, and stood first and second at the show in '64. The second pair were sold there for £32 each, and left a clear butchers' profit of £8 on the biggest; and the first pair returned to Ross-shire, and averaged the next Christmas about £6 more, and 156st. of 8lbs. Mr. Tew does not depend on very high feeding for his success, but he begins with them early, makes an especial point of keeping them well housed, and always takes them up in the autumn, before they have time to lose condition. So much for the roans of Coul Cottage!

Then we tie our mare to the gate of Allness Church, while we search for the grave of the renowned Sir Hector Monro; and on once more through the snugly-belted garths, past the hill of Novar, whose crest looks like nature's gun-battery, with all the varied fretwork of light and shade, from pine and blooming heather playing rapidly over it. Across the Cromarty Firth is the crofter's side of the Black Isle, and quite a curious mosaic of rotations; and we are at the iron-braced monument of Dingwall at last. We do not envy Mr. Arthur Kinnaird his ride, as at 10 p.m. he takes his seat beside the driver of the Skye "mail" for the night. Our own evening would have been dull enough but for a book-auction, where two rival orators were pitched against each other, turn about, for a quarter of an hour each. Still, we did not carry away a single cattle note, except that the Prelacy of the town, just two hundred and nine years ago, reported the people of Apple Cross for sacrificing a bull to St. Maurie; and we thought that it would be a great ultimate saving of beef if some professed bull breeders of the present day would fall back on the Apple Cross habit.

A ride of three or four miles from Dingwall brought us into the very heart of the crofters of the Black Isle. "*Hae ye ony Gaelic?*" was the eternal response when we asked them anything touching their cattle and crops; but in some instances it was only their way of "moving the previous question." A girl went so far as to inform us that the stirks she had in charge were "*just Geordie and Cornoch;*" and one old man seemed grateful enough in Scotch, when we got off and helped to stir a very large boulder which was bothering him. He sealed his thanks by the tender of his snuff-mull, which we declined, as we had done the Sunday before, when it was handed as usual round the church gallery. To it and toddy we have been alike callous. Ben Wyvis bodes no good in the distance; and the caw of the rooks, which we seemed to have wholly lost since they came sailing home from every corner of Caithness to Barrock, is heard on our right above the ruined tower of Kilcoy. Mr. Murray farms there, and the sight of some capital young bullocks for their months confirmed what we had heard—that he always stands high in the local-cattle classes, along with Major Wardlaw and Mr. Cameron, of Balinakyle. A few miles further, and we pass a sort of old white château, with its gate-posts shrunk from the wall; and then from the trim beechen avenues of Belmaduthy, which seems the very garden of the isle, we once more catch a view of the Moray Firth. Pure-bred boars, Leicesters, and shorthorns have all been great points with Major Wardlaw; and the poultry made a far braver show than any we had seen in Scotland so far. Florist by Hiawatha was the Major's bull in '62; and since he was sold to Mr. Tew, the roan Royal Towneley has reigned in his stead. The former, who is the sire of Forth, took the first prize both at Inverness and Dingwall; and the latter has never been beaten in the district, and has won the Highland Society's local medal.

Few sheep, so to speak, are bred in the Black Isle, which boasts of such equality of temperature from its position that, according to the old native belief (which is overthrown annually), no snow will lie on it. The Middletons of Davidstown are the largest sheep-feeders in the Isle. They do not lamb any ewes, but buy half-bred hoggs at home or in Caithness or the south, and either sell them off turnips that year or as dinmonts. Others breed from cast Cheviots, and half-bred ewes as well; but their general practice is to buy and graze half-bred lambs, and despatch them by the Edinburgh steamer from Cromarty when the turnips are ended, or at least by July. The Isle has pretty good grass, and better turnips; and although oats form at least three-fourths of its white crop, the home supply has to be helped out by importation. Beasts are not bred to any great extent, and these are nearly all shorthorn crosses. The Muir of Ord is the great fat and store exchange. Its principal

G

spring markets of April and May are for selling fat cattle, twos and threes which have been wintered and fed in Ross-shire; and in September and October the winterers muster strong. The markets are held on Lord Lovat's ground, and his lordship draws the market customs. Striking the average for 1863-64, there were 1,843 cattle and 4,547 sheep at each of these great spring markets, and 2,265 cattle and 7,429 sheep at each of the autumn ones.

We hardly got a good general view of the more cultivated part of the Isle till we had left Munlochy, and scaled the hill beyond. Then looking forward instead of backward, we soon espied the long-desired and thickly-wooded Tom-na-Hurich (or the Hill of the Fairies) and Inverness, that town of pure Saxon English and that cheerful portal of the North Highlands. The mare jumped out of the ferry-boat into the new county, striking sparks of fire from every hoof in her hot haste to be done with oars and sails, and we place her only too gladly in the hands of a professor of shoeing, and rely for things in general upon the boots at the Caledonian. We should think he hardly ever took part in those running fights for passengers, which raged with especial fury over the bodies of foreigners throughout the season in the old coaching-days; whereas the waiters seemed less dignified, and much more of the figure for a skirmish. It was a wonderful thing to watch that man taking his station at the hotel door before service on Sunday morning, and acting as consulting counsel to the visitors on the different styles of preachers in the town. When he had run his eye once over them, he seemed to fathom their taste. "*You would like a very rousing gentleman; I've just one to suit you: go to ——;*" or, "*I think I'll send you to the Esteblishment;*" and off they went on their various ways, this side the water or over the water, meekly and nothing doubting.

Inverness has already twice taken its turn in the Highland Society's circuit. The thick, short-legged, and "handsome Belville" was here in 1846 on his " grand tour;" and although that caustic moralist, Anthony Maynard, of Marton-le-Moor, would persist to the day of his death in saying that "he was good enough if you backed his hind quarters into a hedge," the judges at the three national shows of that year saw nothing to compare with him. The curiosities of the show were Lord Cawdor's Old Scotch ewes, a West Highland heifer (in which Tom Thumb and Commodore Nutt might have rejoiced), and a trigenarian Highland mare, suckling her twentieth foal. On the second occasion in '56 the brothers Cruickshank all but swept the board with shorthorns. M'Combie sustained his Paris *prestige* to the full with Hanton, Charlotte, her twelve-year-old dam The Queen who was second to her, the Fair Maid of Perth who was third in the cow class, and his other black Venuses. Lord Southesk's Druid was in his lusty youth; Simson of Blainslie and Collie of Ardgay shared the Leicester prizes;

and Brydon, besides winning the first prize for
Cheviot and black-faced gimmers, only suffered
one defeat in the Cheviot tup classes—at the hands of
Donald Horne of Langwell. Inverness's attempts in
a turf point of view have been much less striking.
In 1823 its race-sheet consisted of two matches, two
races of two, a couple of walks over, and hack and
pony races to follow next day. In 1827, when it
was the scene of The Northern Meeting, it soared
into three days, and eleven events, and among them
a Caithness Plate, a Macaroni Stakes in two classes,
a Cromarty Gold Cup, and an Isle of Skye Plate
(given by Lord Macdonald and M'Cleod of M'Cleod);
but only one horse arrived in 1830, and the *Racing
Calendar* knew it no more.

Its character market is the great bucolic glory of
Inverness. The Fort William market existed before;
but the Sutherland and Caithness men, who sold
about 14,000 sheep and 15,000 stones of wool annually as far back as 1816, did not care to go there.
They dealt with regular customers year after year,
and roving wool staplers with no regular connexion
went about and notified their arrival on the church
door. Patrick Sellar, "the agent for the Sutherland Association," saw exactly that some great caucus
of buyers and sellers was needed at a more central
spot, and on February 27th, 1817, that meeting of
the clans was held at Inverness, which brought the
fair into being. Huddersfield, Wakefield, Halifax,
Burnley, Aberdeen, and Elgin signified that their

leading merchants were favourable and ready to attend. Sutherland, Caithness, Wester Ross, Skye, the Orkneys, Harris, and Lewes were represented at the meeting; Baillie Anderson also "would state with confidence that the market was approved of by William Chisholm, Esq., of Chisholm, and James Laidlaw, tacksman, of Knockfin," and so the matter was settled for ever and aye, and *The Courier* and *The Morning Chronicle* were the London advertising media. This Highland wool parliament was originally held on the third Thursday of June, but now it begins on the second Thursday of July, and lasts till the Saturday; and Argyleshire, Nairnshire, Morayshire, and High Aberdeenshire have gradually joined in. The plane stanes in front of the Caledonian (then Bennett's Hotel) have always been the scene of the bargains, which are most truly based on the broad-stone of honour. Not a sheep or fleece is to be seen, and the buyer of the year before gets the first offer of the cast or clip. The previous proving and public character of the different flocks are the purchasers' guide far more than the sellers' description. There is a good deal of caution and sparring on the Thursday, but after dinner on Friday prices begin to settle.

The lots are not at the buyers' risk till they are lifted, drawn, and taken to the road. It is generally bargained to shoot two to the clad score, but drawing properly so as to make the lots really level implies liberty to shoot three or four. The largest lots of ewes or wedders consist of from 1,200 to 1,800 each, and those not disposed of are generally consigned to salesmen for Falkirk, where some of the ewe lots arrive not as cast, but as "double-milled," after having been kept a year, and bringing up a half-bred lamb in Caithness or Morayshire. Some salesmen will "gie the grip" for as many as 15,000; and Gibbons and Lamb of Liverpol, Wallbank of Keighley, Ruddick of Berwick, Nelson, Pattison and Bowstead of Cumberland, Murray and Swan of Edinburgh, Martindale of Manchester, Young of Penrith, and others from Dumfriesshire and Wigtownshire do a very large business. Many farmers also go with commissions from neighbours, and buy and divide a large lot. At one time there were more black-faces in the market, and the brothers Scott have taken as many as four thousand of them, pretty nearly from one farm. Twenty thousand cast ewes and wedders with the Scott brand on them would be travelling south in September and October, by Kingussie and Dalwhinnie, to the Lothians, Cumberland, Dumfriesshire, and the two Falkirks. The brothers would have nearly as many wedders and wedder hoggs on turnips, many of them bred on their own Ross-shire farms, at Fisherfield and Strathnalig, and in the spring of each year the wedders would be told off to the different markets, and the hoggs go back to the farms. Getting turnips for them was sometimes no easy task, and once they were obliged to purchase more than ten score beasts to go as straw-treaders into the East Lothian, as the condition precedent to receiving the wedders.

The new street from the station has given great life to Inverness as far as appearance goes. We like best to be there a day or two before "the twelfth," when the shooters are arriving, and holding mysterious confabs with stalwart keepers, and when some of the youthful hands are on parade in bran-new knickerbockers, and intensely pleased with their legs. It is always High 'Change at Hugh Snowie's on these days, and in fact throughout the season. There sits the veteran behind his shop-desk, with a large file of letters about moors and deer forests to let, at his side —both learning and detailing the latest *on dits* of the trigger. The crack deer heads of the previous season keep, in conformity with custom, their silent vigil of knights on his walls for twelve months and a day. His henchman, Colin Read, has turned out three-and-twenty annual sets of about six dozen each, and he is still working in his little laboratory behind. The tameless eye and the defiant snort of the forest-kings fade into very sober prose while they are put through his crucible. Some were waiting their turn most ignominiously, in boxes full of alum and shavings; the skins of others had reached the higher stage of soaking in a preparation of arsenic, while the jaw-bones were seething in a cauldron; and tow, tin, and putty as props for the mouth completed the *post-mortem* picture. A vigorous spring followed by a summer drought, as in '64, is of course sadly against the grass, and therefore fatal to the growth of the heads; but still Mr. Snowie had plenty of bright antler memories to fall back upon. The Reay Forest in Sutherland has furnished his finest specimen of Caber Slatches, or antlers without tines. A head, which he once prepared for Mr. Campbell of Menzies, had, it is true, only fourteen tines, but the length, span, and thickness of the antlers (which were all covered with indentations, as if the stags had gnawed their comrade while he lay dead for a day or two in the forest) earned a special mention from him. Perhaps the most remarkable of Colin's handiwork is in the Marquis of Londonderry's collection; it has eighteen tines, and each of the brow antlers is not only double the usual length, but, after shooting in front of the head, bends backwards towards the neck.

A 29lb. trout, caught on the Ness, but, as Mr. Snowie says, "*not in a very out-and-out sort of way,*" takes our eye among the eagles and ospreys in the front shop, and reminds us that we might as well take a stroll up to the Loch. It was, however, a bootless errand, as we saw nothing but some very beautiful rock tints. We had no earthly intention of wandering off to Lochiel and Lochaber, to see if the Cheviot is holding his ground after "the Siberian sixty," and we were quite ready to take it for granted that on the top of Ben Nevis there is not vegetation enough even for a black-face. So we halted a day or two at Laggan Cottage, and when we had noted the wooden model of "My First Fish," and all the other paraphernalia of "the merry fisher's" life, we were quite enabled to appreciate the devotion of another of Sir Joseph Hawley's guests (Lieut. Col. the Hon. Fane Keane), who lingered, heroically dead to all St. Leger joys, four days after that party broke up, simply because a large fish was known to be in the Laggan Pool, and landed it triumphantly after a grand struggle of an hour and five minutes. It proved to be of 33 lbs. weight, and the largest fish that had been killed in the Ness within living memory. It is no slight illustration of the beneficial effect of the new Salmon Act, in increasing the size of the fish, that, in the following year, Mr. Denison landed one of 29 lbs., and Captain Vivian another of 27½ lbs. in the same water.

But we had no hours of idleness on hand, and once more we are on the road in a pouring rain, and bid a respectful good-bye to "the lozenge-stone," as we ride through Inverness, away towards Culloden Moor — a table-land on the top of a hill, part of which is improved, while the rest will hardly give cattle a living. Nairnshire, which we entered a few miles further on, after leaving Hill Head to the left, is a lighter and earlier soil than Inverness-shire, and produces an average crop of grass, which is too often sadly cut up during a protracted drought. It has a dry bottom, capital for sheep on turnips and for barley crops, but not very generous for beasts. A large acreage of turnips is taken here for hogging by farmers from the North and West Highlands; and for this reason Aberdeen yellows and globes are more cultivated than swedes, which flourish better in it than all the turnip tribes. The soil is so well suited for folding that turnips are generally at a premium. There is no home breeding to any great extent, and what there is resolves itself into putting Leicesters to half-bred and Cheviot ewes, and selling off the produce when weaned, or holding them ontill they are two shears.

Earl Cawdor keeps about thirty ewes of that old small Scotch breed of sheep, with horns and brown-chesnut faces and legs, which have been gradually pushed aside or "improved away." They are quite the oldest variety in the North of Scotland, and are to be found along the west coast and the small islands from Islay northwards, where they are called "natives." There are some in the Carse of Fendon on the east coast near Tain, but only in a state of keery serfdom. A small flock holds its own near Duncan's Bay in Caithness, and Mr. M'Arthur of Broomtown near Auldearn has also a few, and interchanges with his lordship. The Cawdor flock was formed thirty years ago from a few specimens picked up among the crofters in the carses of Ardersier and Delnies, and others from the late Mr. Anderson of Drumbain, whose flock is now extinct. His lordship's factor, Mr. Stables, considers them "pretty nearly the same breed as the Welsh and allied to the Shetland. The most marked difference is in the length of the tail, which in the old Scotch is seven inches. Add three inches, and they would be Welsh; deduct three, and they would be Shetland." They are handsome and small, much tamer than the black-face, and nearly as good mutton, with fine bone and good wool, which will average 3½ lbs. on a four-year-old wedder. Until that age they are hardly ripe for the knife, and cut up to about 13 lbs. a quarter.

An Angus bull, with some bronze, dun, and black calves, suggests a study worthy of Gourlay Steell; but the "heavy black" is only a foretaste of what we are to find at Lochdhu and Kildrummie, which are respectively owned and rented by Mr. Anderson. He farms 1,200 acres in the counties of Nairnshire and Morayshire, and lets half his turnips in the former for sheep. Mr. Anderson keeps three small stocks of pure-bred cattle, poll, shorthorn, and West Highland, and his tariff for the bull-calves of each breed varies from £10 to £25, £20 to £30, and £12 to £20. He breeds very few West Highlands, and crosses some of the cows with his shorthorn bull. This capital cross is very seldom taken the reverse way, and if a second cross of shorthorn does more for the milk, it does not help the beef. "Hieland humlies," or the cross between the poll and the West Highland, are still found in the higher districts of Morayshire and Bamffshire, and some few on the heights of Aberdeenshire. Their sires and dams have generally been inferior to begin with, and then stunted in their growth, and the offspring are no friends to the grazier. If they come without horns they command better prices, just as a West Highland "down horn" is thought to feed kindlier; but, as M'Combie says of the Galloways, "*they are sad sluggish dogs to feed,*" and very slow waxers. If they have a brown back, they are worse by three degrees, and buyers hate to look at them. Taking even the better sorts, the grazier would make only £1 out of them for every £2 out of a well-bred black or crossbred, but still some of the lowlanders take them in preference to the West Highland. The pick of Mr. Anderson's stock are at Kildrummie, which has also earned a name for Clydesdales and pigs of the Prince Albert breed. Privy Seal, one of the five "Seal" bull-calves, all by Lord Privy Seal, and headed by Great Seal at the Hill Head sale in '60, is in residence as chief of the shorthorn division. This goodlooking roan is a winner both in Morayshire and Nairnshire, and was just fresh from a seven-guineas victory on Elgin Green.

To Mr. Smith's exertions at Hill Head shorthorn breeding and crossing in these parts is very deeply indebted. That neck of land close by Fort St. George, half of it light loam on clay, and the other half sharp and sandy, was not a promising home for shorthorns. Before Mr. Smith took to it in 1841, the Government had been in the habit of letting it yearly, and it had been pretty well scourged in consequence; while heaps of stones and whin bushes did not add to its attractions. The new lessee began to drain on the Deanston principle; but others were so set in their belief that water comes in at the top of a drain, that he faltered in his first idea, and only went two feet deep. He found out his error before the first field was finished, and eventually showed, in the very teeth of prophecy, that swedes and the softer turnips could all flourish there. The foundation of the herd was chiefly laid from Keir, Mr. Cator's, and Athelstaneford; and Lord Privy Seal of the Prince Consort's and Goldsmith of Sittyton breeding left the best mark. Two-thirds of the lots went south of Nairnshire, and among them Lord Privy Seal (150 gs.), who was then a two-year-old, to Lord Kinnaird; and his son Great Seal, and Northern Belle to the late Mr. Balfour of Whittingham. The average of the forty-two lots, which ranged from eight years to as many days, was £35 14s. Neither Mr. Fraser of Brackla nor Mr. M'Lennan was in the list of buyers. The former is well known in the county as a great feeder of cross-bred cattle; and Aberdeen, Elgin, and Inverness butchers all muster round the Meikle Urchany ring, when sixty or seventy cross-bred two-year-old stots and heifers, which have been reared there or bought in at the end of September and wintered very highly, are put up for roup in April.

It was long after dark when we started from Nairn to Forres; and we felt no "pricking of the thumbs" as we passed the spot where Macbeth met the witches. We took the bearings of it, both by road and rail, from the first man we met in the outskirts of Forres. His Shakesperean annotations were on this wise: "*It's just a sort of eminence: all firs and plough now: you paid a toll near it;*" and again: "*I'm thinking—it's just a mile Wast from Brodie Station.*"

CHAPTER VI.
FORRES TO FOCHABERS.

"And find at night a sheltering cave,
Where waters flow, and wild winds rave
By Bonny Castle Gordon."

The haunts of St. John—Morayshire Sellers—"Horned Beasts"—Lack of Breeders—Feeding Stock—Mutton for London—Sheep Farming—Morayshire Feeders—The Forres Fat Show—Altyre Stock—The Balnaferry Herd—From Forres to Elgin—Pluskardene Abbey Bargains—Wester Alves and Ardgay—The Story of the Buchan Hero—Elgin—The Spey Fisheries—The late Duke of Richmond—Gordon Castle—Its Breed of Setters—Its Flocks and Shorthorn Herds.

WE are among the richer wheat lands, and breathe the bright and bracing air of Morayshire at last. The coast to our left is indented with bays and lakes: "The bason of the Findhorn, the resort of innumerable wild-fowl; the sand-hills of Culbin, so curious, almost so marvellous; the Black Forest, stretching away behind Brodie and Dalvey; the 'Old Bar,' where the seals love to sun themselves; the mouth of the Muckleburn, the favourite haunt of the otter," all made the county dear to the heart of a naturalist and sportsman like St. John. But such pleasures are not for us. We can only ride on and remember sadly that he is at rest on a Southern shore, and that the skull of his faithful retriever "Leo" is with him in his coffin.

Morayshire grows all cereals well, on its fine, sandy loam, and the heart of its farming commences with Dyke parish on the banks of the Findhorn, and extends to the Spey. It is the peculiarity of its pastures that they improve beasts faster than any in the north of Scotland for the first six weeks of the grass; but if the season becomes dry, farmers have a very hard time of it, and there is quite a Curtian gulf between Midsummer and the turnips. It would require such a breadth of tares to carry on in July and August, that feeders have sometimes very little option but to sell off or starve on, and naturally choose the former.

"*Take care of your money with the Morayshire men*" is a true saying, as they are dear and hard sellers: and M'Combie has to burnish up his armour when he comes among them. They are gradually extending their grass, and turning from white crops to spotted beasts, of which thirty years ago there was hardly one in the county. Latterly they have been going pretty deeply into sheep; and thus the turnip land levy on the manure-heap increases, while that for the wheat season lessens. The Morayshire "horned beasts," which had a slight cross of West Highland in them, have quite vanished into space. They were chiefly black; and Lord Haddington was so fond of them, that for fifteen years he gave a great lean-stock dealer a commission for 60 or 100 of them at Elgin fair, and would have no others at Tyningham so long as they could be got. Crosses and polls have quite superseded them; but still the general supply falls short of the demand. Farmers complain of the expense of keeping cows, and trust to each other for store beasts to such an extent, that, if a few well-bred ones are found in the market, there is quite a rush at them. Buyers at Elgin cannot echo Colley Cibber's line—

"We triumph most when most the farmer feeds"—

as, before the universal feeding system began, £12 to £13 was thought a fair price for a three-year-old, whereas the same class of beast can now give away the year, and command from £17 to £22, and very good ones still higher, without seeing cake or corn. Some of these two-year-olds are only kept till Christmas, and the feeders look for £1 to 25s. a month out of them on turnips and a little cake.

Five-and-twenty years ago, when a dandy young dealer came to Elgin, and put up at that well-known dealers' resort, the White Horse, the landlady, a large, motherly woman, thus addressed him: "*This is your first visit to Elgin, and you seem a decent young gentleman; now I'll give you one bit of advice —Beware of the Gaulds of Glass and the Cruickshanks of Moray.*" She would have said the same if the shrewd objects of her warning had been sitting in the bar, and no one would have enjoyed it more than themselves. The story merely furnishes a key to the times when two tribes of dealers, numbering about twenty in all, had nearly the whole cattle

H

trade of the North in their hands, and sold to the
M'Combies, the Williamsons or "the Staleys," and
other lean-stock dealers who drove the cattle South.
No men were more respected in their day; and the
same line of business has been taken up by James
M'Donald, the tenant of Blackhillock and Blervie,
who knows every seller for 100 miles North and South
of Forres, and thinks nothing of driving his little
brown horse thirty miles in a morning before others
are stirring. He collects cattle from all points of
the compass, sorts them and lots them, freshens
them up on grass and turnips, and sells them out
again, by tens, twenties, or thirties, to the farmers;
and, in short, nearly three-fourths of the cattle may
be said to go through his hands.

Breeding has not extended much, but the bulls
and the keep are better; and a white bull, as in England, has no special welcome. The breeders have not
adopted the system, which once obtained to some extent in Banffshire, of " getting a calf for nothing"
from two-year-olds. They do not consider that the
calf pays for the deterioration in the dam, which
must be kept yeld for one more year, in order to
recover its handling and prove well at the block, and
even then the London butchers look suspiciously at
the dugs. The stores are generally drawn from Ross-shire and Caithness, but Yorkshire calves, which are
bought at five to seven months, and sold again as
yearlings, have been the sheet-anchor latterly; and
as long as the supply lasts, farmers will not care to

breed. Sometimes they buy them with delivery by rail from £5 7s. 6d. to £7 in October, and finish them off on cake and swedes to the figure of £14 in April. The home-bred calves are generally weaned for harvest time with drumhead cabbage and cake, and so on by degrees through a course of white globes to swedes, with a pound of cake daily till the grass is ready. Most farmers bring up calves by the pail, and consider that they do better in winter than those which have been suckled. Mangels have been all but a failure in Morayshire, and the swedes have been so sure, and advanced so steadily into favour that they now constitute nearly four-fifths of the root crop. Dissolved bones, guano, nitrate of soda, and superphosphate spend another rent in many parts of the county. The lands are top-dressed in March, and the grass is ready for the scythe in May, and twice again during the summer; but the turnips get the greater part of this outlay.

The direct Highland Railway has not only had the effect—except during a snow-storm, when a train was once buried up to the funnel—of curing the truly sublime indifference to time which prevails among the Highlanders, but of opening the London live and dead market in thirty-six hours and 55 minutes—according to the hand-bills. If Aberdeenshire loads its dead-meat train with beef, Morayshire does a good part by mutton as well. We have heard of one great dealer and butcher in

Elgin buying £800 or £900 worth of sheep from one farmer alone, selling the coarser parts about home, and consigning the finer to the South. Most of these sheep were sold at sixteen months out of the wool, and left nearly £2 17s. for keep in their breeders' hands. The light soil suits sheep well, especially when they are folded on turnips; still the cast-ewe system both here and in Banffshire is not of much more than twenty years' date; and before that, part of the ground lay for the cleaning break in naked fallow. Many of the half-bred lambs from Banffshire are spread over Morayshire, as well as Aberdeenshire. Some of these Banff-breds were by Cotswolds; but the shape and size of the head were too often found a fatal objection at lambing, whereas there was nothing of the kind with the Lincoln. Mr. Geddes takes three crops of lambs by a Border tup from his Caithness gimmers, and then turns them off fat; and others are beginning to follow suit. A Shropshire tup on the Cheviot has been a decided mutton success; but Southdowns have never struck root in these counties, either pure or as a cross. The wool of the ewes is 2lb. below that of the Leicesters, and the climate has never just suited them, and in fact the tups seldom command more than 35s. or 40s.

The M'Kessacks, John and Robert, are on opposite sides of the Findhorn, near Forres. The younger brother lives at Balnaferry, and breeds good shorthorns, and feeds largely as well; and the elder winters and finishes off at Grange Green more heavy beasts, polls and crosses, than any man in Morayshire. He numbers from 100 to 150, and now brings out his cracks for the great Christmas show of Forres. Before the establishment of this show in '63, he generally sold his best to Mr. M'Combie, and a lot of two-year-olds at £33 last Christmas was their latest deal. Mr. Harris, who lives at Earn Hill, a little nearer the sea than Grange Green, has come very rapidly into the front rank. Other feeders do not bring them in till the end of August, but he keeps his crack beasts in the yards all summer. At the first Forres show, he took the prize for the best bullock; while Mr. John M'Kessack had it for the best pair of heifers, and cows, as well as for the best cow in the yard. The Society have struck out a very clever plan to increase the entries and encourage the attendance of butchers, by giving prizes for the best lot of ten, six, four, two bullocks, &c., both in the three and two-year-old classes. On the other hand, they have neglected what ought to be the cardinal rule of all Christmas shows—" No knife, no prize"; and hence Venus 9th and Ariadne returned to their pastures, like their English sisters, Victoria, Soldier's Bride, Rosette, and Empress of Hindostan before them.

In both the three-year and two-year-old classes at the show in '64, Mr. Harris had the best single ox, one of them bred by Mr. Garland of Ardlethen, and the other a purchase from Mr. Adam of Ranna. The three-year-old classes were a sort of peaceful duel between the two brothers-in-law, as they had the competition all to themselves, except in one class, where Mr. Fraser of Brackla, a well-known Nairnshire feeder, was second to Mr. Harris's pair, which contained the crack bullock of the day. It might have been led to the shambles, as within living memory many an ox of far smaller pretensions has been, bedecked with ribbons and preceded or ridden by a piper, seeing that it produced for society eighteen hundred and ninety pounds of beef and 264 pounds of tallow. Four out of the ten in Mr. Harris's winning lot were black polled, one grey and the rest red and white crosses; while the Grange Green ten were all black polled. Then curiously enough, in the lots of six, Grange Green, with its red and white crosses, turned the tables on to the Earnhill polls. Mitchell of Wester Alves, Ross of Hill Head, Ferguson of East Grange, Tew of Teaninich, Smith of Minmore, and Garden of Netherton, were among the winners on this December day, when one hundred and eleven beasts were in the ranks, and the Morayshire Farmers' Club and the Forres Fat Show fairly joined hands with Mr. Hall Maxwell as witness.

Sir A. P. Gordon Cumming of Altyre was also among the prize-takers. The baronet, who resides within three or four miles from Forres, rears about fifty young cattle, West Highland and cross-bred, annually, and feeds off the same number, besides keeping a flock of three hundred half-bred ewes. He began to show during the last two years, at first merely at Elgin and Forres, and then he pushed his way with no small success to the Highland Society, Bingley Hall, and Islington. Thirteen firsts, six seconds, and a third make up his winnings so far. Two of the firsts were for a shorthorn bull and a Leicester tup; and the beautiful forelegs of the roan cross-bred heifer which won at Smithfield were dilated on by the shorthorn men with no small delight.

At Balnaferry, which lies about a mile out of Forres, we found fifteen shorthorns and their calves. Venus 9th, with her fine red frame and grand bosom, has been the queen wherever she has been shown, and there were plenty of lusty youngsters to testify to Privy Seal (18642). The herd is of about ten years' standing, and its earliest patriarch was Mr. Geddes' Randolph (15128), who took his first-class degree at Elgin (which seems the touchstone of bulls in these parts), backed up by Fair Service of Shethin blood, who brought in the reds, and Privy Seal, from Hill Head.

Royal Seal (20750) by Privy Seal from Venus 9th, is now in residence, and so is Fashion of the Booth blood, who was bred by the Duke of Montrose, and commended at the Newcastle Royal. Ariadne resembles Pride of Southwicke in her colour, and owes much of her Knightley style to her Bosquet dam; the three-year-old Lady Elma by Lord Elgin, and so back to Lawson's Chief (another Elgin winner), has been first in her class three summers in succession; and Cowslip by Privy Seal is following hard after her.

We remember an old land-valuer boasting that if you only sent him a pot of earth he could tell " you for a certainty whether it grew good carrots, and guess all the rest." The monks seem to have had quite as high tasting powers, as Elgin and Kinloss Abbeys command a rich wheat district, and Pluskarden is on the verge of Alves, that very Goshen of parishes. Mr. M'Combie followed in their footsteps, not with sandalled shoon, but with plaid and trusty staff; and in John Hutcheon's day he bought seventy to a hundred beasts, mostly fours and fives, out of Kinloss Abbey Yard for thirteen years' running as sure as spring came round. He and the old man knew each other well; and as the latter always opened with an ample margin in hand, it took, on an average, three good days to bring them together. They would retire to sleep on it, and meet in the morning, not a shilling nearer; but no other buyer interfered, and they always settled it at last.

The land after leaving Forres seemed to be principally a vast turnip expanse. Mr. Ferguson, of Grange, is not far from Kinloss Abbey, and is known as perhaps the largest buyer and feeder of lambs in Morayshire. He will buy fully twelve hundred, sell the tops as hoggs, and keep on the rest for a time. His neighbour, Mr. Garden of Netherton, feeds from 30 to 40 beasts; but among them last year were the best pairs of two and three-year-old polled bullocks in the county. Then a delightful ride through the Quarry Woods for a mile and a-half brought us from the Earl of Fife's on to the Earl of Moray's property. His lordship's Wester Alves farm is held by Mr. John Mitchell, and, like the adjoining one of Ardgay, is supposed to have the very finest pasturage North of Aberdeen. The high road goes right through the latter farm, whose richest side is on the North. Mr. M'Combie has drawn several of his choicest black stores from it, and paid as high as £29 for a lot of thirty. Many a Highland Society winner has been trained here; and it was with Fair Maid of Perth, which Mr. Collie bought at the Tillyfour sale for 81 gs., that he beat Mr. M'Combie's own Mayflower for the first cow prize at Edinburgh. Zarah was also a heifer of his breeding; and it was from her, crossed with Black Prince, that Mr. M'Combie bred Kate of Aberdeen, certainly the best calf of any breed that we have ever looked over.

A gay, wild-eyed roan with a white calf brought our eye back to Shorthorns in the pastures at Old Mills, where Mr. Lawson has a nice herd; and in front of us, about a mile away, the "sun shines fair" on the cupolas and warmly-tinted sandstone of the Elgin houses. Hard by it is the celebrated Green, where many a shorthorn Waterloo has been fought, and where buyer and seller have set each other like cocks so often, with " The Cock of the North" to look on. There, too, came " The Farmer's Friend" in his simple guise, like an old soldier, always in time, and with a kind greeting and a pleasant story on his lips. Buchan Hero of the white eyelash had passed away from Mr. Ferguson Simpson's hands before that gentleman took up his residence at Covesea, near Elgin. Hence he never joined the bull ranks on the Green; but he won in a still greater fight at Berwick-on-Tweed, against "the English bulls, the Scotch bulls, and a' the bulls." One of his greatest admirers, who had his eye to a " crank" in the palings on that memorable day, thus describes the contest : "*I lookit, and they drew them, and they sent a vast of them back; again I lookit, and still the Buchan Hero stood at the heed. They had nae doot of him then. A Yorkshireman was varra fond of him. And he wan; and Mr. Simpson selt him to Sir Charles Tempest for two hundred. It was a prood day that for Aberdeenshire and Mr. Simpson.*"

We rode through Elgin without drawing rein. Time was pressing, and we were only just able to admire the thistle on the fountain, to wonder why there should be both a "Batcher Street" and a "Batcher Lane," and to glance from the gaunt-eyed, thin-legged wayfarer who illustrates the psalm over the Alms House door, to the ruined cathedral, where the ivy was shrouding the savage handiwork of "the Wolf of Badenoch." About a mile out of the town are the well-enclosed and highly-cultivated fields of Linkwood, with the snug homestead of Mr. Peter Brown, a younger brother of General Sir George Brown. He holds, along with his son, 1,200 to 1,300 acres of arable land, besides pasture; and at another farm near Rothes they have a herd of from thirty to forty polled cows, which they are crossing with Young Hiawatha, a fifty-guinea purchase from Mr. Geddes. Mr. Brown used to be an exhibitor at the Highland Society, and won, among other prizes at Inverness in 1846, one for the best pair of Angus oxen. The road wound round some curious heather knolls, and the long beech hedges and the Gordon tartan, green with a single yellow stripe, soon showed that Fochabers was nigh. The late Duke of Richmond used to tell with great glee how, when other officers indulged in gaudy papers, he lined his tent at Aldershott with tartan during his stay there with the Sussex militia, and how he proved himself the canny Scot by untacking it and carrying it back to Goodwood with him, to "serve in the next campaign."

But here is the rolling Spey at last! Fifty miles up its stream near Kinrara lies Jane Duchess of Gordon, with those beautiful eyes turned even in death towards the Cairngorm, and those lips for ever mute, whose kiss raised the 92nd regiment, when the Gordon tartan ruled from Fochabers to Fort Augustus, and away to Badenoch and Lochaber. The river was still low, and the farmers were earnestly watching for the deep flush on its sandstone cliffs, which forebodes abundance of rain. It is said that on one side the cliffs run under the Moray Frith to Brora, while Southward they die gradually away into the chain of everlasting hills, Ben Rinness, Ben Aigan (hill of the eagle), Knock More, and Muldearie, up whose bonnie brown sides the plough and the shorthorn are slowly but surely creeping. The salmon fishing on the Spey was once let off, but it has been now for fourteen years in the hands of the Duke The seasons of 1862-63 were widely opposite in their character, but the proceeds only differed by £20. Last year the take was hardly two-thirds of an average, but the great falling off was in the size of the grilse, which seldom reached 5lbs. The fish are brought from Tugnet in spring-carts, and so to Fochabers station, and over the Highland line to London within 24 hours of the net-haul.

Shortly before the late Duke's death in 1860, a new outlet was made to the Spey, but it did not just chime in with the temper of this most rapid and unmanageable of Scottish rivers, and taking a turn eastward, it all but cut away the fishing station at Tugnet. Watching the progress of the works to defend the village of Garmouth and its adjacent port of Kingston gave his Grace almost a daily object for a four-mile drive during his last summer at Gordon Castle. When he had seen Tugnet, he would often go and visit a small steading which he was putting up near the railway station. The tenant only paid £8 a year; but he was an old Peninsular man, and there was the great tie. Many and long were his cracks about old times and comrades with Captain Fife, who has also exchanged his sword for a ploughshare. His Grace quite astonished another old "cannon-ball" of the disstrict, who did not know him by sight, when he asked him to fetch his Sunday waistcoat with the medal on it. The old man could not tell for his life "*how the gentleman kenned I wur theer, and that I wur hit gan down the brae at Orthes: It's true enough. Did ye ever hear the like?*" It was the Duke's earnest care that his tenants should do well, and he latterly loved far more to be among his farm improvements and his Southdowns, than he did to go to Glenfiddich, where he had once been wont to spend nearly half of his three months' stay.

The grounds of Gordon Castle are on what was once termed "The Bog of Gight." The spirit of the bog and its attendant snipes have long since fled the spot, which is rich with spreading limes and planes, let alone "Dr. Johnson's oak," which, by the bye, is an alder. It stands, as a memorial of his visit, at the edge of the garden, whose pleasant terraces and vases were in all their flower glory, and banished our old plague, the stunted bourtree bush most effectually at last. The 800-acre home farm is divided between arable and pasture. None of the blood which bore the "yellow and red cap with golden tassel," when Mus, Ghillie Callum, Red Deer, Refraction, or Officious were led up the Goodwood glen on those glorious July afternoons, is to be found in the castle boxes; and even The Rasher and Hartley Buck have departed. Blood stock has no honour in these parts; but it is different with good cart-horses, and we note with delight, in a capital new range of stables, that the 70 or 80lbs. of harness which they generally carry upon them has been reduced in weight more than a-half. We pass the ventilating barn and the Commended at Stirling Leicester (which was on trial with ten highly-caked Southdown ewe culls), to meet Whipper-in (19139), a well-made, fine-handling bull, but rather deficient in his hocks. This Royal red roan succeeded Prince Arthur (16723) by Booth's Lord of the Valley (14837), who was a twin, bred by Mr. Ambler, and left a number of dark reds behind him.

The cows and heifers had to be sought in the park, and we beguiled the way by a chat with Jubb, the head keeper, whose seven-and-thirty black-and-white tans were spreading themselves out like a fan in the kennel meadow. Three-and-thirty of them were just starting to "The Glen," or Glenfiddich and Blackwater, which march with each other. Originally the Gordon setters were all black and tan; and Lord F. G. Halliburton's Sweep, Admiral Wemyss' Pilot, Major Douglas's Racket, Lord Breadalbane's Tom, and other great craftsmen of the breed were of that colour. Now all the setters in the castle kennel are entirely black and white, with a little tan on the toes, muzzle, root of the tail, and round the eyes. The late Duke of Gordon liked it, as it was both gayer and not so difficult to back on the hill-side as the dark-coloured. They are light in frame and merry workers; and, as Jubb says, "better put up half-a-dozen birds than make a false point." The composite colour was produced by using black-and-tan dogs to black-and-white bitches; and at the sale in July, 1836, eleven setters averaged 36 gs. The five-year-old Duke, a black-and-tan, fetched two guineas below that sum. He was bought from Captain Barclay, and begot another Duke still more famous than himself, from Helen. She was also the dam of Young Regent, a black, white, and tan, which joined the Bretby kennel at 72 gs.; and his lordship did not grudge 60 gs. for Crop, although one of her ears had been gnawed off in puppyhood by a ferret. Lord Lovat's, Sir A. G. Gordon's and Captain Gordon's of Cluny dogs have been the only crosses used for some time past at Gordon Castle. Sailor's beautiful scull caught us at once, and Jubb might well say that "he knows everything." Dash lay dignified and apart during the revels, and there was no passing by Young Dash and the neat Princess by Rock from Belle. A dozen pups by a dog of Lord Lovat's, also of the Gordon Castle breed, were out at quarters drawing nurture from terriers and collies.

From the setters we passed on to a half-grown litter of deer-hound puppies, some of them rather too light in colour for the hill, and not of the orthodox badger-grey of Gruin ("Hold him"), who was keeping company with a bloodhound, and three foxhounds, which enjoy roe deer amazingly after their Wiltshire toils. There were one or two retrievers, which brought back the story of the country keeper, who whispered confidentially to a friend at Tattersall's, "*I intend to buy them three dogs if I give a fi-pun-note for them,*" and saw them knocked down at precisely the same average as the setters. A dozen terriers of all breeds and shapes composed "The Lower House" near the ivy-covered water-mill. They lie in their tubs, watching for the rats as they come saucily up to their very trenchers, and then make some fine field practice. We never saw a more motley lot. Snap with his chocolate nose, Jack the rough and ready, Nettle with her Landseer head, Chloe of the turnspit legs, Toby of the prick ear, Gipsey, Spicy, the tailless Peg and Punch, the game black Wasp, and little Dandy, which scarcely weighs 5lbs., and yet makes it a rule of life not to pass a day without killing something.

And on we went to the sheep and shorthorns, past the front of the house facing the park, which recalls Badminton so strongly that we quite expected Tom Clark and the hounds to appear among the trees in the distance, on their way to "the lawn meet." The days of foxhunting at the castle departed with Duke Alexander, who died in 1827. The herd numbers from sixty to seventy, and has been gradually built up since 1842—under the present factor, Mr. Balmer, and his father—by Mons. Vestris (6220), Bloomsbury (9972), Magnum Bonum (13277) by Matadore, Willis's Water King (13980), and Whipper-in (19139). Among the cows, Victoria and Flirt, a capital breeder and milker, both testify to Magnum Bonum; and Princess, the dam of Victoria (whose daughters Duchess 3rd and Mangosteen are the best representatives of Prince Arthur), goes back to Barclay's Pacha. The roan Princess Royal and the light red Mysie are bred on one side from the Shethin herd, and Crown Princess is a purchase from Hill Head. The bull-calves and a few heifers are sold along with the shearling rams and some of the draft ewes every October. There are 350 Leicester and 160 Southdown ewes in the flock, but 60 of the latter are crossed with the Leicester. Three or four years since the Southdowns sold well, but the park (which suits the Leicester to a nicety) is rather too low for them, and the *Filaria hamata* and other intestinal parasites have made far more havoc with their lambs. The Leicesters were laid in thirty years ago from Burgess and Buckley, with a slight infusion from Robertson of Ladykirk. Since then more Border ewes have been purchased, and Pawlett, Wiley, Sanday, and Cresswell tups have all been used; and about eighty shearling tups are sold every year at an average of from £4 to £5. There were some very good Leicester and Southdown ewes in "the Ward," where we sought refuge from the heat under "the Duchess's lime," which spreads out its branches like a banyan-tree. It needed but the drowsy tinkle of the sheep-bell to persuade us that we were in Sussex once more—when Charles XII'ths and Hyllus's match was on every lip—and strolling over, out of mere boyish curiosity, from Bognor to Goodwood, to see the sheep-shearing in the park, and to dine in the tennis-court, and that the Duke, his friend Bishop Gilbert, and Archdeacon Manning with that grand bald head, were "still the first in the throng."

CHAPTER VII.
FOCHABERS TO SITTYTON.

"Take spade and mattock, dig thyself—
Boor's labour makes thee strong and stout;
And herds of many a golden calf
Shall freely from the soil spring out."

A ride from Fochabers to Aberdeen—Back to Orbliston—A Shooting box—The Mulben Herd—The Pig Trade of Banffshire—The Portsoy Cartsire Stud—Clydesdales—Banffshire Shorthorn Beginnings—The Rettie Herd—Mr. Rannie's Leicester Flock—The Montbletton Herd—The late Mr. Grant Duff—His Catalogue Notes—The Forglen Breed of Cows—Mr. Lumsden's Herefords—Hereford Crosses—The Kinnellar Herd—The Sittyton Herd—Udny and Jamie Fleeman.

THE country from Fochabers to Aberdeen is varied enough. For four or five miles the road winds through the pine woods of the Altash Hills, behind which are the home shootings of Gordon Castle. Careful hedgerows and good steadings mark the Duke's property, which goes up nearly to Keith, where the land becomes colder, and acre after acre is without a fence, and occupied by a number of small holders on what is called "the runrig system." Keith is famous for its artificial manures, and we thought of the witches' cauldron—foot of horse and horn of ram in this case—as we pored over the remarkable bone-heap in Messrs. Kynoch's yard. The good grazing begins again about Rothiemay, and on to Huntly, where you descend "Bogie-street," and then scale some fine breezy uplands, with heavy bullocks and eternal oats and turnips all round you. Three white horses came dashing down a lane with such verve that they looked for the instant like Herring's Horses of the Sun. As we neared the "Aughton Forty Daughs" (eight-and-forty ploughs), the black-faces began to dot the policies, and we marked the Castle of Balwhino on the winding banks of the Ury. At Kinethmount the soil is poorer; and then we swept down nearer Inverury upon the glorious grazing of the Garioch district, which had just carried off the Aberdeen Fat Cup. There were skeletons of fair booths on Inverury Green, and children enough to make a Malthus faint. Inverury has a bran new Town Hall, where Mr. Grant Duff, M.P., was about to declare himself at half-past seven p.m.; and the head inn at Kintore could boast of a bronze and knightly sign. Fifty blue painted ploughs were awaiting their autumn labours a little out of the town; and beyond the hill of Tarrybeg we caught a peep of the sea once more, and the very mile-stones changed to granite in honour of *Bon Accord*.

This is the route if you scorn the rail, and ride to catch the mail-train for the South at Aberdeen; but we were not done with the Fochabers district; and in due time we were once more at the Spey side. We strolled down under a burning sun to Orbliston, and found Mr. Geddes, appropriately enough, among his Caithness gimmers. He has been secretary of the Morayshire Farmers' Club (which not long since reached its grand climacteric) for upwards of thirteen years, and has fought and won many a hard shorthorn battle on its Green. Mr. Hunter, of Dipple, who holds the next farm, has a large flock of half-bred sheep, and took a prize with the best crossbred heifer at Birmingham in 1863. Westertown is also not more than a mile from Orbliston, and its tenant, Mr. George Brown, has been well up in the polled classes at several Highland Society's shows. He not only bred "Mulben's" Prince of Wales, but picked Windsor (221) for £40 as a calf at Tillyfour, and sold him for £18 to Lord Southesk, who never repented of his bargain. Mr. Geddes's shorthorn probation spreads over more than a quarter of a century. His first bull was from Chrisp of Doddington, and since then a bull of Colonel Cradock's blood, and others from Nicol Milne, Douglas, Hunt of Thornington, Turnbull of Bowhill, Grant Duff, Lovemore of Rettie, Hay of Shethin, the Cruickshanks, and Stirling of Keir have found their way into his herd-book. Douglas's Duke of Leinster, Turnbull's Hassan (12995), and Stirling's Hiawatha (14705) left some sterling traces behind them, the latter more especially in his cross-bred stock, one of which bore its own testimony in the boxes. Young Hiawatha of his own breeding, and British King (19352) by Lord Raglan (13244), and both of them winners in the district, were there; and when we had peeped into every nook and corner of the capital new bothy, where each man has a separate bedroom, our route was once more down the Spey, on whose banks we found Mr. Wainman and a couple of friends in the snuggest of fishing-cottages. People who read up during the show season must fancy that he lives in a perpetual shower of prize pig telegrams from Fisher, but they are very far wrong. He thinks very little of King Cube, Happy Link, Silver Age, and all the rest of them, in comparison with the lemon-and-whites in his kennel, and the treasures of that strangely curved box for the fishing-rods, which is peculiar to "the throw on the Spey."

It was far too hot for the hill, and we were Banff-bound that night; and on we went by rail through a deep gorge of whin and heather, and passed, a few miles on this side of Keith, the well-known farm of Mulben. Mr. Paterson began in '45, and keeps about twenty Angus cows and queys. His best tribe is the Mayflower, which is descended from a quey of Mr. Thurburn's of Drum, and he has also had a slice from the herds of Lord Southesk, M'Combie, Bowie, Walker, George Brown, and Davidson of Inchmarlow. Black Jock carried a silver medal as the best polled bull in the Banff showyard in '54; and Mayflower (614) was born the next spring, and eventually bloomed into a two-year-old second, and the first cow at the Highland Society. "Mulben" has also had his full share, more especially with bulls, of the Highland

Society prizes at Aberdeen, Huntly, Elgin, Aberlour and Dufftown, and other district shows. In '61 Malcolm (269) took two firsts and a second at the Royal Northern, the Highland Society at Perth, and Elgin; and when Spey, Fiddoch, and Avonside combined at Craigellachie in '63, his Prince of Wales was the best of any breed. Still Mr. Bowie's Tom was too much for him in a capital class at Kelso; but he beat Mr. M'Combie's Rifleman (325) at Aberdeen the next year, and was also first at Stirling in quite a congress of "Princes," to which Sir James Burnett and Mr. Hepburn contributed. However, he was parted with in January last, and his son Sultan reigns in his stead.

Pig information became rather more lively at Banff. They had spoken mysteriously in Caithness of the "old pig of the country," with long ears and body to correspond, and told us how Berkshires, Windsors, and small Yorkshires, and the large black breed or "Neapolitans," which Sir George Dunbar got from Mr. Wetherell, had gradually worked it out. A Chinese pig of Mr. John Wilson's helped matters here, and so did some boars from Kingcausie. This county is the pork centre north of the Frith of Forth, and pigs from Aberdeen (whose Lunatic Asylum fed them to such a size that the neighbours declared they were "*as big as stirks*") are all sent up to it for curing and export. The small white cross from England suits best, and carcases from seven to eight stone of 14lbs. neat find the readiest sale to the exporters. Pork curers from Huntly and Turriff buy them up at that weight; and although a few pigs are sent by the steamer from Lossie Mouth, the majority of them are cut up in quarters, and packed for London in kits broad at the bottom and narrow at the top. In fact, so much is sent away, that the bacon used in Banffshire is generally Yorkshire-fed.

Horse-breeding is also well looked after. We heard of the chesnut presence of The General, whom we last saw cantering away from a large handicap field at Doncaster; of Criffell, a bay trotter, and of a three-fourths bred carriage sire. Portsoy is the cart sire "Rawcliffe" of Banffshire, and Mr. Wilson's great ambition is to combine the Black Comet and Emulator strains. The latter was a chesnut, and he and the grey Remarkable (whose powers of draught in the hands of an Edinburgh contractor quite justified his name) were the only foals that Mr. Wilson retained when he had used a Lincolnshire grey from the Ellon country to thirteen of his mares. He kept Emulator till he was sixteen, and Eclipse, a son of his from a pure Clydesdale mare, is still in his stud. Black Comet was by Little Sampson by Sam Clark's Muckle Sampson, whom Lord Kintore bought at Coldstream and named "Coldstream Lad." Mr. Wilson's show circuit takes in the Royal Northern and most of the local shows in the counties where his horses are stationed for the season, and his grey Comet won the fifty guineas at Glasgow, which was given under the condition that the winner was to travel in Argyleshire for a season. This grey, with the handsome quarters and fine swinging step, was one of the five which were resting from their summer toils at Portsoy. With him were Eclipse, a short-backed black; Robie Burns, rather the smallest of the lot; Tom of Lincoln; and Inkerman, a well-ribbed brown with a Roman nose, and the first-prize honours at Inverness on his head. Horses are very cheap about here; and two very fair ones can be bought for the price of a strong five-year-old yoke ox. On strong land the latter plough best, and we thought how shocked Mr. Atherton would have been if he had seen a Cherry Duke bull at Mr. Wilson's thus earning his daily rations—and not as the penalty of over-fatness or sluggishness—with a bullock by his side.

The cause of the decline of cart-horse breeding in the neighbourhood was the high price paid for large-sized ones, when railways became general in the South. Farmers were tempted by the price to part with their best mares and fillies, and the size and stamp have never been recovered. It is the lorry system which keeps up the Clydesdale size so much, as, if they were eighteen-hand Magogs, they would be greedily sought after. The heels and tips enable them to drag such enormous weights; and in this respect the Edinburgh horses have the advantage over the London ones, who lack them, and are all shod on the late Professor Coleman's principle, that the pressure on the frog was essential to the health of the foot, in order to prevent navicular disease. Professor Dick holds, on the contrary that navicular disease arises in the first instance from a strain of the tendon in the navicular bursa, and is not in any way connected with the shoeing. In Edinburgh, you will sometimes see a horse with three tons on a lorry; and an old blackhorse of seventeen hands once drew a printing-press, which weighed with the lorry above five tons, three miles on the rise, all the way from Granton to Catherine-street. Black is as common a colour as any for Clydesdales. Many of the breed are rather small and sour in the eye as well as flat in the rib; and side-bones, feet flatness, and weak heels are rather common among them. Professor Dick considers that most of their ailments arise from too long yokings and fastings; and colic, distension, and rupture of the bowels are the natural results of gorging at meal-times.

Banffshire has long had yearnings after shorthorns. Mr. William Robertson, of Stoney Ley, got some cows and a bull direct from Holland; but they were big and rough, and when crossed with the common cows of the country, the coats became papery and the flesh light. It is upwards of forty years since Jerry, a massive white, came as a present to the Rev. Mr. Douglas of Ellon, and worked a great reform. He was bred by Rennie of Phantassie, whose array of white bulls had made quite a sensation when the Highland Society first met at Edinburgh in '27. The late Mr. Wilson, who was factor to Lord Seafield at

Cullen House, had been working up to '29 with horned black polls, blacks and whites, brindles, and various other local variations on the doddies, and crossing them with a West Highland bull. The crosses, according to the veteran Longmore of Rettie, were *as good as I ever saw go before their ain tails*"; but still Mr. Wilson's nephew, the present occupant of Portsoy, was quite shaken when he saw Mr. Thomson of Fife's roan shorthorn bull "Comet" at the Highland Society's show at Perth, and he determined not to leave without him. It is a curious proof how little was known of the breed, that the man who brought him to Banffshire persuaded his new guardian that he would eat nothing but oatmeal-porridge and milk; and that a Highland judge gave him a prize at Cornhill the next year *" because I never saw the like of him before."* Mr. Longmore would not fall into the new fashion at first, but he soon came round, and sealed his allegiance by buying a white bull descended from Jerry.

The Bank of Boyne is called "The Egg of Banffshire," and, as far as shorthorns go, Rettie is the yoke of the egg. In '34, its tenant bought Charlotte, a prize cow at the Aberdeen show from Mr. Deacon Milne, and paid Mr. Grant Duff 80 gs. for Jacob (6101) by Holkar (4041). He also got Dannecker (7049) and some queys from the latter, and strengthened his herd from Ladykirk as well. Rosamond, a seventy-guinea purchase, came from Ury in calf to Balmoral (9220) by The Pacha (7612), along with Legacy by The Pacha, whose Balmoral bull calf Inheritor (13065) *"laid me in well."* What with him, Seafield (9616) by Duplicate Duke (6952), and Earl of Aberdeen (12800) from Hay's of Shethin, bull medals began to come in fast at Rettie and there are now six or seven in array on the table.

Imperial Rome (16292) by Lord Raglan from Imperial Cherry was bought from Mr. Douglas as a calf, and begot Viceroy (19054), another of the medallists. Benedict Balco (14159), who was then out at hire, was included in this 280-guinea purchase from Athelstaneford; but he "brought the pleura with him as a compliment" from the train, and it swept down nearly £3,000 worth of stock at a blow. The two bulls weathered it out, and eventually left a large number of heifers behind, which have been crossed with Sir Charles the Second from Erminstade, an 101-guinea selection at the second Babraham sale.

We took a great fancy to a red bullock with a regular Rose of Summer horn, by Imperial Rome from a polled heifer, and but for Mr. Martin's roan stopping the way it would have been first in the cross-bred class at both the great English Fat Shows last Christmas. Tiptree, one of the many bull-calves which Mr. Longmore has sold to Australia, was the sire of a rare heifer of the same cross which was first in her class at Birmingham and London in '60. The herd contains about forty cows and heifers, among which we spent a very pleasant hour, and ended with Mimi, the dam of Viceroy, who was not on the scale of some of the others, but still very bloodlike in her looks. Fourteen to sixteen bull-calves and a few heifers are sold annually, and go chiefly into the neighbourhood or abroad.

Mr. Rannie of Mill of Boyndie, about a mile from Banff, is also a bull breeder, but only keeps about ten cows and heifers. They are descended principally from Red Rosebud, bred by Mr. Grant Duff, and in lineal descent from Holkar, Young Alice, grand-daughter of Alice of Ury, and bought from Mr. Morison of Mountcoffer, and Maid of Judah, one of the heifers at the Longmore annual sale of '54. He has principally used Omar Pasha, grandson of Jemmy (11611) from a Van Dunck cow, and Prince Imperial, a combination of Mr. Longmore's Inheritor and Earl of Aberdeen blood. It is, however, upon his Leicesters that he has taken a more decided stand, and his farm, of which 130 acres out of 581 are grass, is remarkably well suited for them. There are about 200 pure Leicester ewes, and the remainder half-breds. The Leicester flock was commenced nearly fifteen years since by Mr. Rannie and his late uncle, from the late Duke of Richmond's, Mr. Morison's of Bognie, and other flocks. Wiley and Sanday tups were used at first, but the size gradually fell off, and the big and hardy borderers from Chrisp and Cockburn were called in to the rescue. The third prize for ewes fell to

Mr. Rannie at the Highland Society's Show at Perth in '61; and in the aged ram class at Kelso he separated two of the crack Border breeders—Purvis of Burnfoot and Stark of Mellendean—with a very neat sheep, which struck us as very good in the forequarter and round the heart, but small by the side of its two rivals. In fact, we never saw three prize sheep which differed so widely in their styles.

Mr. Robert Walker's of Montbletton is another county stronghold of the blacks. For a quarter of a century he has bred this class of cattle, but did not pay very strict attention to pedigrees until 1850. He has about twenty breeding cows, and has carried all save one of the medals given by the Banffshire and Turriff District Association. Most of his best stock are after The Earl (291), and Tam O'Shanter by Hanton (228). The former took a first prize for him at the Highland Society's third Edinburgh Show; and "Tam," who was bred by Mr. M'Combie, won the yearling prize at Perth, and eight in the district as well. He has not long been slaughtered, and his sons Sambo and Black Diamond, both of them winners, have been used since. Sambo was sold to the Hon. Col. Pennant last January, to cross Welsh cows, and Duke of Cornwall by Tam O'Shanter from Mayflower (614) has been brought forward. Mr. Walker's winnings have not been confined to bulls, as his Mayflower, which was transferred to Tillyfour for 60 gs. at the last Montbletton sale, was first in her class in "Tam's" year at Perth; and his Topsy, which combines The Earl and Tam O'Shanter blood, after beating everything at the local shows, carried first honours in a large class of two-year-old heifers at Stirling, and brought her first herd offering, a bull-calf by Sambo, last New Year's Day.

The history of the late Mr. Grant Duff's mind on shorthorn breeding may be read through the notes of his annual catalogue, which was published every October. Some breeders keep them bound up, and take to them at intervals for light reading on a winter's evening. In point of candour, he was a perfect Mechi, and showed all the ardour of a missal hunter in routing among old breeds. For instance, in one of his sale catalogues, which is headed by "Forlorn Hope, a Shetland cow, 8 yrs. 8 ms.," we have lot 2, "a brindled horn cow or Aberdeenshire shorthorn." Upon her he observes: "The breed is now rare; tradition ascribes their origin to a cross between the Dutch and Falkland breeds introduced by the laird of Udny; they have great properties as milkers and feeders." However, she only made £12; and "the brindled polled cow of the old Forglen breed, which gave milk while unbulled," beat her. Then we have lot 4, "a grey and black mixture horned breed from 'the auld town of Carnousie breed'." His grieve, Mr. William Jamieson, considers that the Forglen were generally yellow polled, superior as dairy cows to the polls, and very kindly feeders. In fact, the district claimed to put them quite on a par in flesh with the West Highlander. The breed was at the Forglen home-farm when the Ogilvies were the lairds, and began as follows. The last Lord Banff's mother procured two Devonshire cows and one bull, whose produce were for some time kept pure. They were crossed with the native horned breed, and then with the Aberdeen polls, and from them came the fine yellow cows, known as the Forglen breed. Shortly after the late Sir Robert Abercromby came to reside at Forglen, he commenced breeding from the Eden herd, and nearly twenty years of crossing has made the old yellow breed hardly distinguishable from pure shorthorns. The lady who introduced the Devons was great-grandmother to the present baronet, who inherits much of her taste for cattle and other improvements.

What seemed to many the mere enthusiasm of yesterday in the late Mr. Grant Duff has proved the wisdom of to-day. He quite rises into prophecy in some of his foot-notes, when he utters a warning voice against overfeeding for shows, disregard of pedigree, and careless crossing. A cross-bred bull was his aversion; and he gave it as his experience that, although you could not perhaps do great harm by putting a Shorthorn West-Highland bull to a poll or a Poll-Shorthorn cow, the union of similar crosses never succeeds. He never wearied of proclaiming the virtue of that West-Highland cross with a shorthorn of which breeders are now seeing the full value, both as regards flesh and constitution. One cross of it was all he went for, and hence his remark that "people have asked me for West Highland crosses in defiance of warning." He fully allowed that "a good beast is a good beast however come;" and then he adds, most wisely, "but we cannot depend upon succession without pedigrees." Upon every point he took the public into his confidence, and gave copious reasons for his new belief. When he abandoned his prejudice against stock by Lord Kintore's bull, it was "the stock of old Rose that compelled me;" and when he began to wean his calves on oilcake, he only did it "in deference to contrary opinions."

His reverence for the English "shorthorn homes" and their owners was unbounded. Marquise, a fifty-guinea heifer from Sir Thomas Cartwright's, seems to have been one of his earliest purchases from them, and she proved cheap at ninety to Mr. Longmore. Alice was bought at Mr. Charge's sale in the May of '45, and he tells with no small glee how at the Yorkshire and Durham County she had beaten Mr. Booth's Bud or Modish. Then we learn how "the grieve was sent for particular information, and perhaps to buy one of Sylph's descendants," and how he came back with Ladye Love (whose dam Belinda departed to found a tribe at Babraham) for £67 and expenses, but not in-calf. Again did that devoted grieve cross the Border "expressly to bring back Carnation by Benjamin (1710), dam by Ganthorpe, and so to

Foggathorpe;" but it was upon the tribe of Brawith Bud, "that special legacy of Peter Consitt to Wilson," that his best herd affections were set. It was a great story of his, and never out of the foot-notes on any consideration, that "300 gs. had been offered Mr. Wilson in his presence for her daughter Carmine, who weighed 98 st. imperial as a yearling at Thirsk, if he would only guarantee a calf."

There was no resisting such Mrs. Armitage proportions, and she came to Eden as a speculative bargain for £77 16s. 7d., expenses included, so that the public were put right on that point to a penny. Brawith Bud was not long in following her for 170 gs.; and he had the delight of vanquishing at the Brawith sale both J. Booth and J. Maynard, the latter of whom "was heard generally to say that he still thinks her one of the best shorthorns in England." This was the great cow purchase of his life, and he was luckily enabled to record of her that she had paid him 100 per cent., was useful till eighteen years of age, and "never a moment unwell, except for a few days in 1848 from epizootic influences, or rather epidemic influenza," which, as he afterwards observes, "only confirmed the health and vigour of the patients." She was of a vigorous, longlived sort, as her sire Sir Walter was good at sixteen; and her daughter Jenny Lind, from whom sprang the Kirkhee sort at Sittyton, was rejected along with Mint by Robin-a-Day at Turriff, because their long hair gave the judges an idea that they were crosses from West Highlands.

The Messrs. Cruickshank often bought from him, and there was a good deal of joking, when Premium was knocked down to them, with the foot-note "It is more than probable her next will be a bull-calf," and it proved to be a heifer.

Holkar, bred by Mr. Bates, was a great fancy of his, and he delighted to write of him and Sir Fairfax 2nd as "the rival bulls." He was so worked up with a desire to possess him that he offered 500 gs. for him if twenty-two bulls and bull-calves by him reached that sum at a sale. They did not, and he got his wish gratified at a much easier rate. Still, Robin-a-Day was his "Comet" of the North, and he gave the Formartine district more especially a proof of his good "Carcase" descent. The price was 44 gs., and both he and Mr. Knox made more than 200 gs. a-piece out of him. We have lingered very fondly over these relics of one who blended so much hearty enthusiasm with his science. Unhappily for us, he had died ten years before we looked on the woods of Eden. "Jenny Lind, 100 gs. (Mr. Tanqueray)," was the highest lot at the roup, and the Messrs. Cruickshank marked their estimate of Brawith Bud with 92 gs. for her Pure Gold.

We bade good-bye to Banff, and got out at the Turriff station for a six-mile walk in search of the only Hereford colony in Scotland, which lies about six miles from the rail, through the heart of the old "Kintore country," that "Nimrod" speaks of, in his delightful "Northern Tour," as "one in which more enjoyment of hounds may be had than in any other that I saw in Scotland." It held a very fine scent, and many of its best gorses were made by the late Earl. His " huntsman's stall," as he termed it, was at Gask, where he rented a farm, and built kennels and stables at his own expense. Of hounds, horses, and hunting, he was a rare judge, and never were servants better mounted. After his death, Mr. Urquhart kept about eighteen couple, and hunted part of the country for three seasons, and since he gave up, the note of a foxhound has not been heard in it. The country in some parts made us half-fancy that we were near Ashdown, looking out for the coursers; but the scarlet of Mr. Nightingale is seen there no longer, and the once great Turriff Club has sunk into a very minor affair among a few farmers.

Auchry, with its swans and islets on the lake, and the old-fashioned manor-house, quite aided the Wiltshire illusion. Mr. Lumsden owns about four thousand acres, and has reclaimed fully a fourth of them from waste and heather. Of this he has laid down about seven hundred acres in permanent pasture, and roups it out annually. He uses Border and Leicester tups on half-bred ewes, and his black Essex pigs are from Sexton. At the outset he crossed Ury shorthorn bulls with Aberdeenshire cows, and sold the produce to the butcher at three years old; but twenty years ago he was smitten with a fancy for the Herefords, and has never since wavered. His argument is that they have a thicker

K 2

coat of soft hair to stand a northern climate, that they can be made ripe for the shambles on mere grass and turnips without extras, and that the cows after the calves are weaned can do well enough on oat straw and water, and thus save the turnips for the rest of the stock. He prefers crossing the Hereford bull with the Aberdeen "cross-bred" cows, and considers that the produce do not lose their aptitude for fattening, but grow to a larger size than the purebred. His favourite instance on this head is of a four-year-old steer, which, according to the London butcher's certificate, weighed 1,919 lbs. neat; but still he finds that, as two-year-olds and weighing from 7 to 8 cwt., they leave the most profitable return. They have oat straw and turnips till Christmas, and then, we believe, about 3 lbs. of cake per diem till May, when they are sent off to the London market.

His first venture comprised a bull and two cows from Mr. Hewer, and two more from other Hereford breeders. Conqueror, the bull, cost 60 gs., and had three horns. By way of making good his claim shortly after his arrival, he fought a shorthorn and smashed the surplus horn, which was seven or eight inches long. Some years after Mr. Price's celebrated Sir David (349) was purchased for 100 gs., and stayed there three seasons, when Mr. Turner of Noke Court arrived and brought him back to England.

As a proof of the goodness of the cross, he once sent a shorthorn cow, with her calf, yearling, two-year-old, and three-year-old, all by a Hereford bull, to a show at Aberdeen. Deacon Milne bought the heifer yearling, and said that its beef was so fat that he was "obliged to send it to London, where they'll eat anything." The Deacon also bought the two-year-old, and its weight, which decided a wager, was 8 cwt. 1 st. and 4 lbs. neat, while the three-year-old after six months' more keep realized £45. Well might the late Duke of Richmond observe, as he passed down the ranks and scanned the family party, "*That cow owes her owner nothing.*" Mr. Lumsden has lately had a visit from a very kindred soul, Mr. Duckham of the Hereford Herd Book, who bought two heifers for him at the Westonbury sale, and also sold him his bull Cato (1902) for the purpose of stamping the Sir David character on his herd; but so far the white and mottle faces have made very little progress, and small farmers too often use neither Hereford nor shorthorn, but some wretched cross with a light body and long legs.

A few years before Mr. Lumsden began, Mr. Mitchell of Fiddenbegg got a Hereford bull and two queys, and imported much more recently another bull in partnership with his cousin at Haddo. Mr. Shepherd of Shethin had also a Hereford bull, but they are gradually giving it up. At Fiddenbegg (which we must take a little out of its order) we found one substantial trace of the system in a cross-bred bull, which leant a good deal to the Hereford. He seemed a rare thriver, and was most freely bellowing his dissent from the short commons they were obliged to put him on. A few cows in the herd had retained some of the Hereford colour, and others merely a little white under the jowl and round the eye. One, in which a Hereford strain on each side had united, was pretty nearly a pure Hereford to look at, and another of a similar cross quite as decidedly short-horn.

Thirty miles down the rail brought us to Kinaldie station, two or three miles beyond Kintore. Mr. Milne of Kinaldie, a well-known breeder, has a herd of shorthorns within a very few minutes' walk of the station, but we had not the good fortune to meet with him. Kinnellar is on the opposite side of the rail, and about a mile up the hill. Its tenant, Mr. Campbell, began with shorthorns from Ury eighteen years ago, and Isabella by The Pacha ((7612), for which he gave 20 gs. at one of the Captain's roups, was the first that did him substantial service. She bred a calf soon after she was two, and followed it up with eight more, and carried the Highland Society's prize at the Marr Association, as the best beast on the ground. Ruby Hill, from the Hill Head sale, also did well for him, and Miss Ramsden by Duke (3630), Nonpareil by Lord Sackville (13247), Thes-lonica by Duke of Clarence (9040), Crocus by Sir Arthur (12072), and Thalia by Earl of Aberdeen (18200), have all been good breeders. With the exception of Lord Scarboro' (9064), a purchase at Mr. Wetherell's sale in '59, all his recent bulls,

Mosstrooper (11827), Beeswing (12456), Garioch Boy (13384), Scarlet Velvet (16916), and Diphthong (17601), have been bought at Sittyton. Beeswing and Garioch Boy, the latter of whom he considers his best, were both by Matadore (11800). "The Boy" died very early, and never could be shown for the Aberdeen Challenge Cup, which Mr. Campbell won with Scarlet Velvet and Diphthong in 1862-63, and then resigned without a struggle to Forth. Scarlet Velvet went to Morayshire, and so to the block; and Diphthong, a very thick, good bull, with a curious mark like a tape-line round his left fore ribs, was never beaten till he came into that grand class of 21 aged bulls at Stirling, and stood fourth after a very close finish to Van Tromp, Fosco, and the Worcester Royal winner Duke of Tyne. There are about forty cows and heifers in all at Kinnellar, and Prince of Worcester (20597), bred by Mr. Fletcher, has been used of late with Diphthong 3rd. Fifteen to sixteen bull-calves are sold every February, and the top prices in 1863-64 respectively were, Diphthong 2nd 101 gs. (Mr. J. Ross), and The Provost 75 gs. (Mr. J. Suttie).

We then committed ourselves to the guidance of a fifty-nine page catalogue, with red edges, containing 216 females, from Pure Gold to The Gem, and learnt from it that Sittyton was thirteen miles north of Aberdeen, and within three miles of Newmachar Station on the Formartine and Buchan line. The scenery about Newmachar is rather wild and bare;

but when you are once fairly among the beech hedges and deep woods of Straloch, where a whole colony of rooks

"Find a perch and dormitory too,"

the chesnut mare with her foal at the corner of the copse, the ivied bridge, and the keeper with his pointers, make up rather a pretty " bit."

The brothers Cruickshank, Amos and Anthony, devoted their earliest Sittyton energies to the old Aberdeen poll, under the orthodox county belief that it would grow larger and ripen earlier than any " beef-cylinder" north of the Tweed, but still their shorthorn beginning dates from '37. They went to work in a very cautious way, with a cow in-calf, called "Durham Countess," but short in the pedigree withal. Her first produce was a white bull, and then, after rather a discouraging two years' interval, came a roan heifer, Peeress by Barclay's Sovereign, of Mason's Lady Sarah tribe. The latter was second in the cow class, when the Highland Society met at Aberdeen; but, although she then brought the maiden premium to Sittyton, she never had a heifer calf. Of a trio of red heifers which came next from Mr. Smith of Elkington, near Louth, only one, Princess by Lowdham (10477) ever bred; but Moss Rose by Grazier (1085) and Carnation by A-la-Mode (725) turned the scale, and their tribes remain to this day. The white Inkhorn (6091) was bought from Captain Barclay as a cross, and was used for two seasons; when Premier (6308), another of Lady Sarah's (" the stang of my trump," as the Captain termed her, and a 150-guinea purchase at Mason's sale), was bought to replace him. This was also the figure which Messrs. Cruickshank subsequently gave for Fairfax Royal (6987), of Mr. Torr's breeding, at the Walkeringham sale, and the Premier and Inkhorn heifers, for whom *he* was destined, amounted with others to about twenty-five.

The Ury sale in '38 added largely to the herd, and Clara by Mahomed (6170), her heifer Barcliana by The Pacha (7612), and Strawberry by Second Duke of Northumberland (3646) were among the accessions, and all founders of tribes. Fairfax Royal's stock turned out well, and his son Prince Edward Fairfax from Princess was used for a season, till Velvet Jacket (10998), a first-prize winner at the Highland Society and at Aberdeen, took his place for another, and then departed across the Channel to Mr. Latouche. Matadore (11800) introduced Booth's Hopewell blood, and won wherever he was shown, and after doing good Sittyton service in the shape of a fine harvest of roans, he was living, in his sixteenth year, at Mr. Allan Pollok's in Ireland, and was even a prize-winner at thirteen. His son Lord Sackville (13247), from Barcliana by The Pacha, went on till he was in his sixth season, and got some capital cows, but nothing has made his mark more profusely and decidedly on the herd than The Baron (18833) by Baron Warlaby, from Bon Bon of the Sylph tribe.

He left upwards of two hundred calves, and at one time forty-five females in the herd were by him, principally from Matadore cows. Mr. Cruickshank only bid twice and got him for 155 gs., at Mr. Tanqueray's sale. Master Butterfly 2nd (14918) had quite as fine a chance during the twelvemonth and a day that he survived his 400-guinea purchase at Bushey Grove, but he failed to improve it, and finished up by dying of affection of the brain. His stock were generally red, and Cherry Bell and Clementine the best of them. John Bull (11618) and Lancaster Comet (11663) (the sire of Champion of England (17526), whose stock are coming out well in hair and flesh) have also been on the bull list; and Lord Raglan (13244), Ivanhoe (14735), Lord Garlies (14819) by Heir-at-Law, Mr. Peel's Malachite (18313), Mr. Ambler's Windsor Augustus (19157), (the third yearling bull at Battersea), Forth (17866), and Sir James the Rose (15290) have been brought Northwards in turn.

Females were gradually bought up at every shorthorn sale; and Watson's, Wilkinson's, Grant Duff's, Holmes's, Robinson's of Burton, Speerman's, Dudding's, Hopper's, Cartwright's, Wood's, Marjoribanks's, Towneley's, Fawkes's, &c., have all contributed something to the grand Sittyton total. The largest sale at one time has been to Mr. Marjoribanks, who took ten heifers in a lot at 1,000 gs.; one of them Khirkee, the founder of a most profitable tree. This was followed by the sale of five more, with Pro-bono-publico by Matadore as their esquire, to Earl Clancarty, in Ireland : but now, as a general thing, the brothers only sell a few heifer calves to run out the less valuable tribes, and keep about thirty to supply gaps. They have had as many as 108 calves, and have sold about 80 in one season, as they very rarely make bullocks of them. The annual bull roup was commenced in 1842 with seven or eight lots, and has been continued every October since. About five-and-thirty bull calves, varying from five to nine months, are generally disposed of at it, and five-and-twenty more are the subject of private bargain. At the roup, which is held on the first or second Thursday in October, £44 8s. for twenty-four in 1861 is so far the highest average. Fairfax Hero, Magnum Bonum, King of Sardinia, Challenge Cup, and Conqueror the top prices in 1847, '53, '56, '61, and '64 averaged within a shilling of £100 ; and Magnum Bonum still leads at £115 10s. Vine Dresser and Lord Aberdeen, which were sold privately within those periods, would, curiously enough, have brought up the average to the same point less a shilling, for seven. Some go to Ireland and the Orkneys, and the rest are scattered over all the northern counties from Caithness to Forfar.

The calves always run with their dams from four to five months if they are heifers, and rather longer, on an average, if they are bulls. Except at Udny Castle there is no old pasture land on the three farms; and the whole of the cows and heifers are obliged to be tied up for six months of the year, and kept on turnips and oat straw. Showing is not much cultivated, and training for shows still less; and, except at the Royal Northern at Aberdeen, the Highland Society, and very occasionally at the Royal English, the herd makes no public appearances. Whenever it does, there is a good old Sittyton rule, that the show cattle must earn a clean bill of health for nearly eight weeks, and a field at the Longside Farm is specially devoted to them.

The brothers reckon upon about a dozen leading tribes. The Fancy tribé from Captain Barclay crossed well with Matadore and The Baron; and their Orange Blossom produce was both prolific and profitable. From Sunflower of Ladykirk descent came neat but small bulls; Lord Spencer's Sibyl nicked best with The Baron; and Verdant from Chrisp's of Doddington with The Baron and Plantagenet. The Venuses of Rennie of Phantassie's blood showed a great aptitude for crossing with the polls; Secret had the Bates quality to begin with, and Lord Sackville supplied the substance and improved the head; Duchess of Gloucester, a daughter of Tortworth Chance, brought heifers large in the frame and great in the milk vein, and the bulls of the tribe are after Moir of Tarty's own heart. Nothing laid the foundation of more prize-winners, or hit better all round, than Captain Barclay's Strawberry; and from Wilkinson's Lancaster tribe came Lancaster 25th, which was sold with five cows and heifers in a lot at 100 to 150 gs. each to Mr. Barclay of Keevil, and calved the Lord Raglan triplet.

"The Champion" was at Udny farm, but there were still five or six left for a grand parade at Sittyton. "Forth" was roused from his lair, and Anthony Cruickshank took him in hand. Nothing will reduce that wonderfully level bulk; and even when he had been starved for some weeks, to make him take to turnips, his pipe of oaten straw quite solaced him. He was terribly sea-sick both ways on his journey from Aberdeen to Newcastle (to which he got a fifteen shillings return ticket), and yet his eye was as bright and his handling as firm as ever, after a Sunday's rest in the show-yard. In short, there never was such a philosophic "fat boy in Pickwick." The Czar and Lord Chamberlain, two sons of Lord Raglan, were in the hands of the men; and Lord Raglan, once nearly as victorious in his *crusade* to the show-yard as his mighty sire beloved of Anthony Maynard, came roaring in the old style round the corner, and kept Amos and his white crush-hat almost on the run. Still it seemed very doubtful whether he would live to see the turnips; and Edward Cruikshank's charge, Sir James the Rose, once the great prize heifer getter of Athelstaneford, was only a grand ruin. There were ten bulls, five here and five at the other farms; and Grand Monarque, Lord Chamberlain, Sir Walter Scott, and Prince Imperial have taken the place of these two seniors, to carry on the great object which their owners have held to so steadily for nearly thirty years—"flesh production for the people at large."

The Smythy Park disclosed a still greener old age in Barcliana, *æt.* 16; and there too was the venerable Hawthorn Blossom. Seventeen bull calves were grazing in the Top pasture, and all getting their cake twice a day for the October sale. The Corn Yard Park, Well Park, The Naggon, and Lead Park had also, in the words of an author whose name we don't remember, lots of "veal just tottering on the verge of beef." As for wheat, four wet seasons had pretty nigh washed it out of the five-shift rotation in these parts. In turnips, the brothers hold faithfully to swedes and Aberdeen yellows (purple and green top); and the potatoes they "rejoice in" are Irish Cups, with their white and purple flowers, as being especially free from disease.

On our right, as we ride to Clyne, and stretching away over rather a low-fenced level to the sea, are the Formartine East and Buchan districts. Forty years ago, when Marr of Cairnbrogie was in his zenith, "Timberless Buchan" was great in "doddies"; but the Shorthorn has stealthily invested its twenty square miles, and the roans and reds have all but worked out the original blacks and brindles.

The Hill pasture at Clyne was full of heifer calves; and we looked with some interest at Golden Days, the last of the eight Pure Golds. Then we drove on for three miles past the church at White Rashes to

Longside, in whose thatched outbuildings on the hill Malachite was once in residence. There was an old "castle" near, but it is a delusion and a snare when you get to it, and it is all converted into stalls. However, if our antiquarian yearnings were quenched, we found solace in the Great Field, where twelve of the prize cattle were undergoing strict quarantine. The road to Udny past Cairnfeckel and Pitrichie was less bleak than the one we had traversed, and crosses of Shorthorn on Aberdeen were grazing on each side of it. Haymakers were seated in a group under the hay, discussing their brose, and three policemen, with their "minds taen up wi' affairs of the state," formed an equally earnest knot hard by. The fortunes of Udny would have formed rather a mournful theme for them, as the old Castle is desolate, and two owners, father and son, had died within the year. In horse-racing England, the inn-sign would have borne the family arms on one side, and Emilius on the other; but the bay has no root in village memory, as he was never in Scotland, and we even forgot him ourselves on entering the stables, in the contemplation of the jet-black Glasgow collars, with the red tuft, the daintily disposed plough gear, and the well-kept brass on the hames.

"We've 122 here, *calves* and all," said our guide; so we first ascended the tower beloved of pic-nickers, and caught a bird's eye view of the objects of our search, as they roamed in the park and fields below.

Jamie Fleeman, "the laird of Udny's fool," is the real hero of the place, in Scotland's eyes. Where one Scotsman knows how "Buckle steered Emilius, at the speed of the express" ten thousand have read how Jamie rode on a stick before my lady into the town of Aberdeen; and his advice to his master to manure a barren soil with factors, " as they always thrive," will stick *in sæcula sæculorum*, when the most golden-mouthed sayings of sage or poet find few lips to quote them. His last words, "*Dinna bury me like a beast,*" went right to the mark as well; and when years had passed by, they were inscribed on a stone, which was placed over his grave. The tower, where he was wont to make sport for lords and ladies gay, now only cumbers the ground. Its huge fire-places are a mockery, and cook nothing nobler than potatoes or brose in a hoer's kettle; and the whole place seemed given over to some faded school decorations. Scufflers, harrows, and ploughs block up the kitchen; and guano-bags are piled in "the state bed-room," which seems about eight feet by four. The planting of the estate is almost as incomprehensible as the brain of its fool. There stands,

> "Like dim cathedral aisles,
> The shade of beechen arches,"

which, starting from no premises, and coming to no conclusions, merely help to make the fields snug for the cattle.

But it was time to quit these great nursery-grounds

of the Caledonian shorthorn. We had arrived at the Six Gates of Udny, that trysting-place of every elf, fairy, and goblin in the country-side; the twilight shadows were fast falling on the avenues of which it is the mystic centre, and we did not care to be caught at the revels.

CHAPTER VIII.

SITTYTON TO ABERDEEN.

"Tillyfour's the place to see
A lot of stunning beefers,
Cows and calves and hummel bulls,
Stots, and likewise heefers.
Every July at our show
He makes a demonstration,
Carries off the medals
And creates a great sensation."

ABERDEEN CHAMELEON.

The Aberdeen Fat Cup—Mr. Stewart's Cracks—The Tarty Herd—Mr. Martin's Show Beasts—Aberdeenshire Feeders—Early Days of Feeding—The Original Horned Doddies—Aberdeen Meat Supplies—The Cattle and Dead Meat Trade with the South—The M'Combie Family—Easter Skene—A Day at Tillyfour.

"Look here! *I never saw so much tallow inside a hare in my life!*" said an Aberdeenshire man to us at dinner. His remark was characteristic of that great feeding county, so especially staunch in its allegiance to King Beef. The struggle among the best bullocks, taking age into consideration, for the Fat Challenge Cup on the links each July is the Derby and St. Leger of the North rolled into one. In '63 it was a quieter affair; but last year we heard, and on all sides, how Conglass's doddy, the first of the breed that ever won it, was the roughest of the three which had that grand finish to themselves; and when we met him at

Islington, we had no doubt on that head. It was on every one's lip how undauntedly Tom Swan had stuck up for Mr. Adams' two-year-old, and how Mr. Moir had the cup dashed from his lips just when he thought his red three-year-old was safe of it, and that it would go to Tarty for ever and aye. This was the fifth year of the Cup, which Mr. Stewart won on the first two occasions, and lost, like Mr. Moir, at the third time of asking.

The leading Aberdeen butchers dearly love to have some choice beast going on for Christmas; and Esselmont, which lies near the railway, a few miles beyond Newmachar, is Mr. Stewart's "training stables." In '63 the roan ox was in residence, and strolling magnificently up and down his yard, clad in a sheet, to keep the pile on for Darlington. He was then in all his four-year-old bloom, and those who only saw him on his last Christmas tour, sixteen months after that, never really knew how good he was. His breeder, Mr. Moir of Tarty, won the Aberdeen Fat Cup with him as a two-year-old, and sold him to Mr. Martin, who sold him to Mr. Stewart for the Darlington Cup contest. After winning it, he was sold and raffled; and the winner of the raffle sold him to Mr. Martin, who trained him for another year, won two head prizes with him at Birmingham and London, and sold him to a butcher for £65. On the sire's side he is in direct descent from Cruickshank's Baron; and for his prize-winning predecessors in the South we had to fall back on our own Christmas recollections, and photographs from Mr. Giblett's paintings.

First came the white ox (by a Shethin bull) which won the Fat Cup at Aberdeen in '61, and at Darlington in '62, besides a prize at Poissy, and the Cup. The red one, which first shook the equanimity of the Durham men, was a very massive fellow, bred by Mr. Garland of Ardlethen; and when Mr. Stewart had marched his "red, white, and roan" on to them successively their Challenge Cup departed over the Border. Two shorthorn cross heifers and a poll brindled ox did good work at London and elsewhere in '61; but '62 was, after all, "the exhibition year" for Esselmont. Fourteen beasts were prepared, and sent off to Darlington, London, Leeds, Birmingham, York, and Liverpool, and when the circuit was over, Mr. Stewart could reckon his prizes by the clad score. This was the year of the red Shorthorn Poll, bred by Mr. Thompson of Dumbreck, which got the Cup as the best beast in the yard at Birmingham, and was only beaten for the same honour in London by a Devon-Shorthorn ox of Mr. Henry Overman's.

Mr. Moir's of Tarty is only four miles from Esselmont in the Formartine district, which is divided from Buchan by the Ythen; and his farm is on the Ellon property belonging to Mr. Gordon. The neat, gabled house and steading are in Tarty Slack, a slight hollow, only two miles from the coast; and thirty cows and heifers, some of them with ten or eleven crosses in them, have the pick of 500 acres. The soil is rather light and sharp, and grows turnips to a great size. Its subsoil is the great secret of successful feeding in Aberdeenshire, and Formartine, with its light yellow loam, and the Vale of Alford dispute the grazing palm. In the former the grass is rather earlier, but Alford has more shelter. With liberal top-dressing cattle can be got into the pastures by May morning all over the county, but May 10th is the average time. This grass is pretty fresh till "St. Partridge day;" but at Tillyfour, the two last weeks of July and the two first of August are quite the best.

Mr. Moir works entirely with "shorthorn crosses." His uncle, who dwelt at Ardlethen, began with Bertram, a Phantassie bull; and the nephew made his first breeding essay at Tarty with a cross of Jerry and Bertram blood. After that, he used Ury bulls, and had a slice of the redoubtable Pacha, as well as of Cruickshank's Fairfax Royal and The Baron. His three-year-olds last year were by Shepherd of Shethin's Red Knight, his twos and his yearlings by Cruickshank's Lord Stanley (16454), and his calves by Shepherd's Earl of Elgin (20170); and Duke of Leeds by Lord Raglan was there, to speak for himself, tethered to a stake under a picturesque shelter of rocks and trees. His calves run for about six or seven months with the cow, and he cuts from twelve to fifteen every year fourteen days after their birth. About September 10th they are brought in and tied up. There are no yards here for yearlings or anything, as the bleak uplands have something more than "a kind of starved look," while the cold blasts come whistling from the German Ocean.

About forty are fed off every year; and Mr. Martin, of Aberdeen, who bought both the roan and red Aberdeen Cup winners of 1862-63, is one of the principal customers. A fine red cow, The Queen, was one of the daintiest quenes of the herd, and the Darlington roan's dam had a very deep-fleshed little "prize-fighter" by its side, to which good colour only had been denied. The herd was in three or four different detachments, principally according to age, and it was quite novel to hear different beasts told off as they came up the brae. It was "half-sister to Darlington ox," "dam of red ox," "half-sister to red ox," "half-brother to red ox," and so on through pages of bullock history.

Mr. Martin's stud was in a long barn at Mr. Fiddes's of Wateridge Muir, not many miles away. There were the red Tarty ox which had just won the Cup at Aberdeen, and a great grey Shorthorn-West Highland ox which stood forcing till the Christmas of the next year, and then took the Liverpool Cup, as the best fat beast. We fancied him more than the red, and well might Peter Allardice, their trainer, observe that *"he fills a string well."* Still, amid all these good ones, Peter could not forget the Esselmont roan, and styled him *"the dashest ox I ever saw."*

With a candour so rare that it deserves mention, he still expressed a doubt whether Mr. M'Combie's big black of '63 should not have *"beaten that red there"* at Aberdeen; and when his Poissy recollections were evoked, he spoke in concise and appropriate rapture of certain Boulogne cookery.

The Formartine, Ellon, Buchan, Alford, and Garioch districts may be said to have about thirty leading feeders, averaging, with the exception of Mr. M'Combie's, from about forty to seventy bullocks each. In old days, when they were entirely dependent on the Aberdeen butchers' custom, four to six was thought a spirited venture. They all breed as well as buy, and finish them off their third Christmas at about £30 a-piece. Doddies are very rare in their lots; and Mr. Bruce of Mill Hill is the only one who has imported the white faces from Hereford fair. Within the last forty years turnips have increased a hundred-fold, and swedes in a still greater ratio since the time of the foot-and-mouth plague in '39. In fact, steam and bone-dust have worked quite a beef revolution. At first only very few turnips were grown on the North-East coast, and the farmers attempted to feed a portion of their cattle. They had, however, attained no proficiency in feeding science, and no method as to the times of feeding or the quantity of turnips to be used; cattle men had double work in consequence, and the beasts grew in years, but not in weight. Drovers came in the spring, and generally took them off to Barnet fair, with the exception of the supply to the Aberdeen market, and the few heavy-weights which went to Glasgow. The cattle generally were about half-fed, and only in fresh store condition. Mr. Anderson of Pitcarry, Kincardineshire (who is still alive and managing his own farm at eighty-five), was the first man who ever sent cattle by steam from Aberdeen, and the first that were ever trucked by rail were a lot of Highlanders belonging to Mr. Hay of Shethin, This gentleman and Mr. Whitehead of Methlick had once a great shorthorn-bull trade, but the latter retired in '56, and when Mr. Hay died, his nephew, Mr. Shepherd, carried on the herd, and sold 140 of them by auction in '63. Shethin has no mean name in shorthorn annals. First Grand Duke (10284) was there for a time as a calf, and was then sold to Mr. Bolden, who sold him to the Americans for 1,000 gs., and Bosquet (14183) and Second Cherry Duke (14265) left their decisive mark.

Red is the fancy colour of the county, and eight-ninths of the beef come from shorthorn crosses, nearly all of which, if their breeding were looked into, could satisfy the "Herd Book" conditions. The cracks are generally picked up as yearlings by the "racing butchers" as quickly as the London horse-dealers descend on to a young hunter in Lincolnshire or the Midlands, and kept by them for 2 or $2\frac{1}{2}$ years for the Aberdeen Fat Cup and the English Christmas prizes. One or two of the best have been Shorthorn-Poll; but this hornless cross is not so common, and Shorthorn-West Highland is as much liked as long as it is kept to the first cross, which will make its £24 easily at two years old.

Black has always been the fashionable "doddy" colour; but brindles, duns, yellows, and greys were once very prevalent, in Aberdeenshire. The yellows especially had a character for early maturity, which was not shared by the brown-back and brown-mouth sort. Many of them had white horns with black points, and as wide-spreading as a West Highlander. In fact, a horned Aberdeenshire was quite as highly thought of as a black poll, and the Huntly· and Strathbogie districts were their especial strongholds. Northamptonshire and Leicestershire men were very fond of them, and some of the leading Cumberland jobbers would occasionally carry away as many as 120 in one lot from the Falkirk October, and winter them in their yards till spring. They were then passed on to Barnet fair, and sold fat out of the Essex marshes in July, August, and September. Old graziers shake their heads mournfully, and say that no such beasts browse the marshes now. Thirty years ago the trade was at its height, and so were the Millers, Billy Brown of Carlisle, Jemmy Reay, and the Temple Sowerby men. The brothers Armstrong, as oral history avers, have been known to go back to Yorkshire from the fairs North of Aberdeen with 600 horned and polled runts, none of them less than three, and generally four years old. Gradually the lots became half horned and half polled; and then the Cumberland men, finding that the horned beasts took up so much room in their yards, tired of them, and turned their attention to sheep and Irish beasts; while the Lothians and Fife, which had once clung very tenaciously to the polls, veered round to shorthorn crosses. The result is that only the "very trash of Aberdeen polls" are to be found at Falkirk now, and a very slight sprinkling of Galloways.

The Aberdeen butchers, who are formed into a guild, and elect a deacon annually, supply themselves in a great measure from Ellon, which has a stock market on the first and third Mondays of every month. Its supplies are chiefly drawn from Buchan and Formartine, and in some of the best spring "markets off turnips" nearly four hundred beasts will be pitched. Old Meldrum has also its fortnightly gathering of crosses from the parts round Udny and Tarves; while polls come thicker at Alford, and are found occasionally at Huntly from the Garioch district. A few sheep are sent to Turriff, Huntly, and Inverury; but the great majority are to be found at Brechin in April, and more especially in June, and at Trinity Muir in the same month. The first "Muir" market is more for fat cattle and two-year-old grazing stock, and so is the April fair at Glesterlaw in Angus. Still, the leading Aberdeen butchers do not depend on these casual supplies, and take grass parks in summer and turnips in the winter, and buy half-breds or blackfaced wedders from the hills to stock them.

Except on a Friday, it is rather difficult to find any of the leading Aberdeen butchers at home. They are always scouring one of the three beef counties— Banffshire, Morayshire, and Aberdeenshire—to look after their beasts in training, or to buy for their shop and the cattle or dead-meat train. The Crimean contracts gave a great spur to the thing in Aberdeen. Forty or fifty extra bullocks were killed every day for the army, and men who went into the carcase business have never left it since. The dead-meat is rather superseding the live-stock trade with the South. Butchers not only send them up cheaper this way, but the hide and tallow are worth quite as much at Aberdeen. Dundee buys the heads and feet, and the tongues, livers, and hearts never go begging at home. In fact, more beasts are slaughtered weekly in Aberdeen than in Glasgow. The butchers kill two and three year old bullocks, queys, barren cows, &c., and dispatch the heaviest supplies from Christmas till the middle of May. In the height of the season the Messrs. Martin will slaughter as many as one hundred sheep and forty beasts, and send very little of it away. Mr. Stewart has also more of a home trade; while Messrs. Butler, Knowles, Skinner, and White send off large live and dead-beef supplies, but comparatively few sheep carcases. The dead-meat train goes at three o'clock, morning and evening, during the season, and the cattle train at one p.m. on Thursdays; and the live stock are pretty equally divided between the steamer and the railway. The latter carries them on the average in thirty-six to forty-two hours, but beasts get more knocked about in the trucks; and the steam companies charge £1 a-head, and are their own insurers to the full value of the cattle as well. Hardly any dead meat leaves Aberdeen by the steamer, and the average in 1861-63 was only 87 tons against 8,943 by rail. In the last-named year, 13,798 head of cattle were sent off from Aberdeen southwards by rail, and of these 9,623 went direct to London. Caithness, Orkney, and Shetlands in that year exported no less than 8,740 cattle, 23,124 sheep, and 844 pigs; and of the 6,000 odd from the two latter places (which are said never to have any disease), two-thirds were landed at Aberdeen, and the rest went on to Granton.

But M'Combie and the polls were still unseen. "The powerful, pushing, and prosperous race" of M'Combies are first heard of in Glenshee and Glenisla*. The name signifies "son of Thomas," and the family is specially mentioned as Clan M'Thomas in the clan-roll. They were all men of large stature; and the "great M'Comie" kept the Cateran in such check, that one of their number thus announced his death: "Blessed be the Virgin Mary! the great M'Comie in the head of the lowlands is dead, for as big and strong as he was." One of his descendants, Donald M'Combie, settled in the North,

* I am indebted for these particulars of the M'Combie family and Easter Skene to a very interesting article in the *Banffshire Journal*.

and an ancestor in direct descent of the cousins at Tillyfour and Easter Skene was buried just 150 years since in the parish of Tough. His grandson William began to farm Lynturk in 1748, and was reputed to be the strongest man in seven parishes round. He had also, like "the big M'Comie," seven sons, of which the three youngest—Thomas, Peter, and Charles—all became lairds. Thomas was an Aberdeen baillie, and left Easter Skene to his son William, the present possessor, who also got Lynturk through his uncle Peter; and Charles, who did not care for the quiet life of an Aberdeen merchant, and preferred the more exciting one of a lean cattle dealer, invested in land, and left Tillyfour and Tullyriach to his eldest son, the Rev. Dr. Charles M'Combie, minister of Lumphanan, who lets them to his brother.

Easter Skene lies midway between the Dee and the Don. Its present owner succeeded to it in '27, and since then he has reclaimed the whole of the estate from heather and bog, and, with the exception of some on the north-east side, has planted every tree on it. The plantations alone extend over 130 acres, and the stone fences to 30 miles, supplemented, whenever shelter is required, by a hedge of beech or hawthorn. Many of the fields have been in grass for nearly twenty years; and when they are broken up, only a single crop of oats is taken, and then turnips. This root is never known to fail, and finger-and-toe is unheard of, which seems to suggest that the disease is rather the result of exhaustion, and the remedy to be found in rest, and not in stimulating manures. The view from the house, a tasteful building in the Elizabethan style, is one of the finest in Aberdeenshire. In the foreground you catch a glimpse of the loch of Skene, and rising just behind or as distant outposts are the Hill of Fair Corrie, of the Birds, Stone of the Mountain, The Fairies' Hill, where, in obedience to the Hogg codicil, a bonfire blazes on the eve of May Day, the Cairn of the Eagle, and the Mountain of the Boat. The huge figure of Morven looms against the Western sky; and Ben Avon, which guards the southern approaches to Banffshire, can be seen best from the farm of Drumstone, where the renowned laird of Drum sat down and made his will ere he strode, claymore in hand, to his doom at the Battle of Harlaw.

The Easter Skene herd is not so numerous as the Tillyfour, but it has held its own right well in the show-yards. It was first in the cow class at Aberdeen, in 1853, with Queen of Scots, beating Lord Southesk's Dora and ten others; and also headed the yearling bull class the same year with Rhoderick Dhu (89). Another of its bulls, Alaster the Second, beat Fox Maule (305) on the same ground, and the only occasion that he was ever beaten. Royal Scot also took a silver medal there; and the ox with which William M'Combie gained the first prize, the last time that the Highland Society met at Glasgow, was born and bred in these pastures.

The pilgrim from Aberdeen to Tillyfour must

keep two great directions, positive and negative, in his head—firstly, change your train at Kintore; and secondly, don't get out at Tillyfourie station, as scores have done before you. That "*i. e.*" is anything but demonstrative in this case. Cluny Castle, which is said to be the finest granite building in Britain, the woods of Monymusk and Fetternear, with Ben-a-Chie towering behind them, are all pleasant landmarks in the twenty-three miles; and the fertile vale of Alford just opens upon you, and gives a bright foretaste of the Braes of Mar, as you leave the train at Whitehouse. Tillyfour is only three miles from this point, but the outlying farms are more easily reached through Alford. The wind was not in the East, and therefore we were promised a dry day at last, and a really fine sight of the vale, which, save Ellon and Tarves, is said to carry bullocks to a greater size than any in the North. Its barley is in especially high favour with brewers and distillers. It suits turnips, both Aberdeen yellows, purple tops, and swedes, remarkably well; but there are no mangels.

Stewart's Inn, to which grouse-shooters and tourists resort in the season, and find no "*purée* of horse-beans," but good hare-soup awaiting them, was our first halt. The entrance-hall is hung about, not with "pikes and guns and bows," but with enormous fox-skins; and it is some consolation, when one thinks of that terrible sacrifice of good fox-flesh, that the landlord sends South all he can get out alive from the hills, and that ten brace, "with black four inches up the pad," were transported to one English county in 1863. Alford owes much to the cottage architecture of its principal proprietor, Mr. Farquharson, whose mansion, as well as Whitehaugh and Forbes Castle, is a leading object to the right, with the heather hills as the glorious back-ground of all.

Still we wanted "hoof and horn" figures in our landscape, and a short ride past the well-filled pastures of Mr. Reid, a successful grazier and prize-taker, and the forge of Mr. Sorly—the "Professor Dick" of the Vale—and so along the banks of the Don (midway between which and the Dee, Tillyfour may be said to lie), brought us to Dorsell's, the first of Mr. M'Combie's four farms. It belongs to Sir Charles Forbes, of Newe and Edinglassie, and consists of about 640 acres equally divided between arable and pasture. Ninety beasts were billeted on it, and when we saw them they had been nearly a month off grass, and had kept up their bloom on tares three parts ripe, which given in this state do not induce scouring, and have much finer feeding properties. Green tares make milk rather than beef, and Mr. M'Combie has long abjured them. The first lot were eating their oat-straw and Aberdeen yellows, and the sheddings were judiciously darkened to encourage digestion and repose. They were all threes-and-fours, and "just good commercial beasts," to adopt "Tillyfour's" favourite term, when he is not especially sweet on anything. Not a two-year-old found a place; as their fore-quarters are seldom good, and their tallow supplies are short. The London butchers have been bitten once too often by them. Three-fourths of the ninety were hornless Aberdeenshire, and the rest blacks with white legs, greys, and reds, brindles, half-bred shorthorns with poll heads, blacks with the loose scur (which is the saving clause of "Doddyism"), blacks with horns pointing one up and the other down, and here and there one with the infallible "mark of the beast" on his buttock, or the real Pagan roan.

The sample grew higher as we proceeded, and reached the Christmas-table candidates for both metropolises, and Liverpool as well. Twenty of them stood in at £23 12s. 6d., twenty-eight at £25 5s., and twenty at £20 10s.—all from Mr. Robert M'Kessack's, of Grange Green, near Forres, while a smaller lot of seventeen came from Dandaleath in Morayshire. On we went through the rest—four blacks together, and very difficult to whip apart; three "heavy Scotch greys," one of which was pretty nearly the head of the lot; and then, close by a red roan with quite Marmaduke crops, stood a spotted monster of full seventeen hands. Mr. M'Combie drily polished off this Magog as "just a heavy beast for shipping," and he was finally sold by Mr. Gibbons at Liverpool for £52. "That completes the eighty," and then came another lot loose in the sheds, ready to take their place in the double stalls as soon as the Christmas beasts have gone. We had not time to go in search of the bull Champion, as it was long past

M

noon, and an October day was not to be trifled with; and with a glance at the beautifully cross-gartered oat-stacks, which stood in platoons four deep, with William Turner, the bailiff, as chief-architect, we once more sped on our way.

There was nothing to take us to the Castle of Craigie Var, whose strong black loam on the granite has furnished Mr. M'Combie with some of his richest pasturage, and we turned off to the "training" quarters at Bridge End. Its 230 acres are rented by Mr. M'Combie from his cousin at Easter Skene, and John Benzies, with his blue blouse and Kilmarnock bonnet, is captain of the depôt. His military decoration is the Dutrone medal presented to him at Poissy, as the servant longest in command of polled cattle; and he has also the Smithfield diploma as the feeder of the best Polled Scot. John is a perfect almanack on the subject of fat shows, which seem to act as milestones on his journey of life. Birmingham and its foundry and factory people have long been a great terror to him. "*They are a dreadful lot,*" he observes, "*with all their pinching and poking; the gentry are very civil, but this gas it punishes the beasts worst of aught.*"

After all his travels by sea and land, he maybe said to have lived so much among the shunts and the breakers (of which his master has given such a vivid catalogue in a recent controversy) that it is only wonderful to see him in the flesh at all. The neat-boned Rifleman by Rob Roy Macgregor (267), from Pride of

Aberdeen, who was first in his class at Battersea, is his peculiar charge; and then came seven dozen bullocks, of which at least seven-and-twenty were "tops," and getting specially sent along upon cake and corn. Three of them, however, were on the reserve list for the next year's shows; or were at least to have the benefit of the doubt, when their companions were dispersed by March into many a British larder.

The whole of the shedding is more useful than ornamental, and heather, tiles, slate, wood, and thatch, all play their part in the roofing. It was here that the great Poissy bullock was fed, and John waxed eloquent at the remembrance of him, although neither he nor his master have heard his weight to this hour. Thrice has John crossed the Channel, left six bullocks behind him and brought back £370, a cup, seven gold medals, two great gold medals, and silver and bronze galore. A Tillyfour ox was second for the Cup, on the first occasion, to the Duke of Beaufort's shorthorn, when Mons. St. Marie's casting vote was said to have decided the day. The Poissy ox, *par excellence*, was bred by Mr. Tough, of Deskie, Aberdeenshire, and sold for £28 at two years old. He was then resold twice; purchased by Mr. M'Combie, for £45, from Mr. Shaw of Bogfairn, kept two years, and finally sold to the Emperor's butcher for £84, after winning £285 in money and cups. Nine-eight was his best girth; and he had this peculiarity, that he would never touch corn.

His training was not unchequered. After Smithfield, John escorted him, along with the Birmingham cup heifer, to Mr. Maydwell's farm in Surrey, and lived with them there till the middle of April. It was not a jovial time, as they brought a Christmas-box along with them in the shape of "foot-and-mouth"; and although the bullock bore up bravely, and only bated an inch, the crack heifer "took off six or seven inches as level as it went on," and had not recovered her bulk when she went to Poissy. Mr. M'Combie first saw her at the Dumfries show, and his mind was not at rest till he had given her breeder (the Duke of Buccleuch) a fifty pound cheque for her, which she returned with interest.

Of the great prize ox of '63, which occupied the box of honour, Mr. M'Combie might well observe, prophetically, that "a little man would not be able to see him without assistance"; and in default of a ladder, John adjured us then and there to mount the manger, and survey (in Athelstaneford phrase) "the vast plateau" of roast beef. "Have you ever looked over more pounds?" was his triumphant query, as we descended. In that low-roofed tabernacle there seemed but one reply. Still the Islington building quite dwarfed him, and we should not have remarked on him as a veritable Great Eastern among the bullocks there. In his leading points he was rather rougher than some we have seen from Tillyfour; but if he lacked the bloodiness and levelness of the Angus, he was, in Benzian phrase, "beef to the root of the lug." He cost £48 at two years and three months old, and was bred by Mr. Stephen, of Conglass. His first prize was won at Garioch; £40 and a gold medal were his two-year-old guerdon at Poissy; and at Liverpool, Aberdeen, and on the "grand tour," he gathered £130 in all. Still, what with some 2,000 miles of travel on his head, and the keep of eleven dozen weeks at ten shillings, there was no such great margin of profit even after a £80 sale. Still he had the honour in his death of being bracketed in point of price, head of the Beef Tripos of the year, with Mr. Heath's gold medal Hereford ox.

Two or three work oxen were being fed off, and laying it on pretty satisfactorily, seeing that flesh has too often a tendency to run to tallow after these furrow gymnastics; but no coaxing could push on the bloodiest black about the place. He was such a beauty that for two years Mr. M'Combie had been at him every way, in and out of the house; but his stomach refused its office, and the tape only told of eight-feet-five, and there he stuck month after month. He would have gone to "some side show" that Christmas, but his level, high-bred form melted his owner, and he kept him on to West Highland years of discretion. His great beauty was his breast and neck vein, but he was rather light in the twist and flank. In the spring he began to take a start, and reached 9 ft. 3 in.; then he went back again, and finally girthed 9 ft. 1 in., and won the head prize both at Birmingham and Islington. By way of set-off to the expenses, the ever wakeful "Tillyfour" sent some turnips in the truck with him, which took a two-guinea prize in Bingley Hall. On the journey he lost very little, and, in fact, it is only the half-trained beasts that go to pieces then; but his appetite failed sadly, and all the fire was out of him when he had gone through the dreaded Birmingham ordeal, although he was so wild at home that he required six beasts and four men to coax him to the station. He was tried with every kind of food; but bran, oatmeal, and barley-meal never exactly suited him, and what he digested best was linseed well steeped in hot water with a little bran and meal in it.

By way of a change, we stepped aside to see one of the cleverest jobbers in the county, not more than two bow-shots from Tillyfour. His white pony was grazing in the meadow below, and the journeys of the pair would fill a ledger. We had seen variety enough that morning, but nothing to be compared to his long byre. A large spotted cow, which cost £1 a leg, was at the far end, next to an eight-guinea Dutch one, and a big black; and two out of the trio made £29 10s. Then there was a cross-bred heifer, hob-nobbing with a black polled yearling, two dun West Highlanders, a brown sunken-backed cow, two polled heifers and a yellow polled ox, and, to crown all, a shorthorn bull with a pedigree and another without. We asked how the former was bred, and his owner responded that he was of the "Viper tribe," and at once produced his printed pedigree from his waistcoat pocket: "Rodney got by Jasper, grandam Viper by Second Billy."

Mr. M'Combie's father bought Tillyfour with the century. Five-twelfths of its 1,200 acres are arable, and one-sixth old grass. Its heights and hollows furnish fine natural shelter, and it is well-watered by burns rising in Bletoch, Tillyriach, and Corannie. It was here that William M'Combie was born in 1805, and learnt that fine experience from his father which has caused him of late years to be regarded both in Great Britain and the continent, like Jonas Webb in another sphere, as quite a grazier king. Mr. M'Combie senior was equally eminent in his business of a lean-cattle dealer; and his son has thus written of his early career, in the *North British Agriculturist*:—"When a young man, he went to the far north to Caithness, Sutherland, Skye, and the islands, and bought large droves of Highland cattle, and brought them home; they were often disposed of by public roup in this county, or driven to the southern markets. At that time there were few regular markets in these counties; but the dealers when they went to the country 'cried a market,' or published that they would meet the sellers on a certain day, and at a convenient place, and in this way the trade was carried out. Large profits were obtained, but the dealers were liable to heavy losses, especially in spring, the cattle being then but skin and bone, and many dying in the transit. My father lost in one night, after swimming the Spey, seventeen old Caithness runts. There were no bridges in those days. It came on a severe frost after the cattle had swam the river. Their bones bleached in the sun on the braes of Auchindown for more than thirty years, and remains of them were visible within the last few years. My father not only carried on a very large trade at the Falkirk markets, but had a very extensive business to England; he kept a salesman who attended all the great English fairs, particularly in Leicestershire, and sold drove after drove that were bought by my father here. Referring to documents in my possession, I find he had in one year 1,500 cattle at the October Falkirk tryst, 900 of which were Highlanders, and the remainder Aberdeen cattle. The Highlanders were grazed in Braemar, on the Geldie, Boynach, and Corryvrone, the property of the Earl of Fife. These were, in fact, his special glens, and the greater part of the £3,500 which he made at Falkirk in two successive years came off them. Prices of cattle were very high at the time of the war. I observe the prices of three heavy lots of horned Aberdeen cattle sold into Cumberland, viz., £22, £23 10s., and £25 a-head. A Carlisle carrier, I have often heard my father say, was the purchaser. He declared that he bought them for eating up the horse litter. Heavy losses were sustained when the peace came. The late well-known George Williamson had a very large drove of cattle in hand when the news of peace arrived, and he was passing through Perth himself with his drove at the time the bells were tolling the merry peal on account of the peace. "Old Staley," as he was called, often said that this merry peal was a sorrowful peal to him, for it cost him £3,000. From my father's books it appears that the expense of travelling was trifling from the north in the end of the last and beginning of the present century. Men's wages were 1s. 6d. a-day, and they received no watching money. There were no toll-bars. The roadsides and the commons afforded the cattle their supply of food.

After his father's death, in 1830, Mr. M'Combie settled at Tillyfour, and followed, until within the last fifteen years, the lean cattle trade to which he was bred, besides keeping a few milch cows and grazing 200 or 300 cattle. There were invariably 60 horned Aberdeenshire beasts among them, which were generally the "tops" at the October Falkirk, and after wintering in Cumberland passed on to Barnet in the spring. As a young man he was fond of coursing, and once won, and again divided the All-aged Stakes at Turriff with Amy of his Buy-a-Broom sort, which he still remembers " as going from the slips like a shot." He also delighted in shooting, and made some very large bags, but his health has been more delicate of late years, and all his field sports have been given up one by one. The Vale of Alford Society was his first showground, and he had not been much more than two years at Tillyfour before he was placed first with a bull which he had purchased from Morayshire. He won again in 1837, and since then he has gradually fallen into the round of the Vale of Alford, the Royal Northern at Aberdeen, and the Highland Society. Inverness and Aberdeen (twice over) have been his greatest weeks with "the Highland," as he swept almost everything in his way; and his blacks were "well on the spot" on the only four occasions—Windsor, Carlisle, Battersea, and Newcastle—that there has been an opening for them at the Royal English. He sent fat beasts to the Birmingham and Smithfield Shows as early as 1840, but it was not until 1859 that he and his black brigade became a leading feature there. During the last six years he has regularly taken the Smithfield first prize for the polled Scot bullock, besides the first in 1861 for the heifers. The latter not only won the gold medal for him as the best female, but took the cup as the best beast in the yard at Birmingham (where his bullock firsts during the same period are only one below Smithfield); and both English and Scotch papers might well unite in their protest, when Mr. Faulkner's shorthorn Dolly, a year older and two inches less in the girth, and by no means a perfect specimen of her kind, was preferred by the Shorthorn, Devon, and Hereford judges in the contest for the Smithfield gold medal to the beautiful "sable interloper."

He laid the corner-stone of his fortunes by the purchase of Queen Mother by Panmure (51) from Mr. Fullerton, then of Mains of Ardovie, near Brechin, and now of Mains of Ardestie, near Dundee. She was then a yearling heifer, and cost but £18 at a cheap time.

As she turned from her few first services, she was put for a penalty to draw wood, and did all the ridging up of thirty acres of turnips as well. She then proved in-calf to Monarch (44) (who was bought by Mr. Ruxton at the Ardestie roup), and the heifer was called after Lola Montes, who was then in the height of her Bavarian conquests. Queen Mother's first prize was at the Vale of Alford. She was then third at Aberdeen, and even with twelve summers on her head, she was good enough not only to take to the Highland Society's meeting at Inverness, but to stand second when she got there to her granddaughter Charlotte*, and to beat Fair Maid of Perth and fifteen more capital cows.

From her the family-tree branches off in three directions, through her daughters Lola Montes and Bloomer by Monarch, and Windsor by Victor (46). The last-named was the dam of Windsor (221) by Hanton (228), who was sold to Mr. George Brown of Westertown as a calf, and was passed over by him to the Earl of Southesk (who was first at Edinburgh with him) for £180.† Crosses for the produce of the Lola Montes and Bloomer lines were found in Hanton by Pat (29), who was purchased at two years old, with a quey, from Mr. Bowie, for £110, and Angus (45) by Second Jock (2), which only cost £36 at Mr. Hugh Watson's roup, and has also done yeoman service to the herd. Angus was used to

---

* By an oversight at page 83 Charlotte is called her daughter; and the cow herself is spoken of by her original name, "The Queen."
† In page 116 this price, owing to the slipping of a figure, was made "£18."

Lola Montes and Bloomer (which, like Windsor, was first at the Highland Society and Windsor Shows), and Charlotte and The Belle, another Highland Society first, but not with the size of her dam, were the respective results. Hanton, whose show career embraced nine firsts from Alford to Poissy, where even the Emperor could not buy him, got both Pride of Aberdeen and Daisy from Charlotte, who also had Crinoline, née White Legs, by Victor 3rd (193); while Fancy was the produce of him and The Belle, and his son Rob Roy M'Gregor (267) followed suit with Lovely. And so the succession has gone on—Monarch, Angus, Hanton, " Rob Roy" (267), Black Prince by "Rob Roy," and lastly Rifleman, who is by " Rob Roy" from Pride of Aberdeen—a son and daughter of Hanton—which is as nearly in-and-in as Mr. M'Combie dares to go, much as he likes the blood. Kinnaird Castle, Balwhyllo, Ardgay, and Monbletton have also furnished their contingents in Empress and Dulcimer, Lady Agnes, Zarah, Mayflower, and another Mayflower, &c. Mr. M'Combie bought both these " Flowers," after they had stood first and second at Perth, and liked the second-prize one decidedly the best of the two.

Scotland is very true to her champions, and when all this thought and energy culminated in the Poissy and Battersea triumphs, four hundred neighbours and breeders, with the late Marquis of Huntly in the chair, assembled to do Mr. M'Combie honour by a banquet, which was one of, if not the largest, ever held at Aberdeen. "The English agriculturists always maintained," said the hero of the evening, "that a Scot would never take a first place in a competition with a Shorthorn, a Hereford, and Devon. I have given them reasons for changing their opinion (deafening cheers)." The old champion, Mr. Hugh Watson, was present for the last time in public, and in a few graceful words he tendered his congratulations, and spoke to the glory of the Angus, whose name no time will sever from his own.

"Black and all black" is the password at Tillyfour, and no roans, greys, or brindles, or beasts of any other livery, are allowed within its lines. The fortress lies on the top of a hill, and the steep ascent terminates at last in a little grove of limes and ashes. Behind is the great sky line of that bare and bleak Forest, which once was Royal Corannie, and away to the right is the Glen of Tillyriach, and that evergreen gorse, which knows no Rallywood challenge. The black-cock often descends from his heather heights, and shares, with about thirty Galloway and Angus yearlings and two-year-olds, the outlying hundred acres of the Nether Hill, to whose rich qualities the perpetual burrow of the blind little "gentleman in black," beloved of the Jacobites, furnishes the highest clue. Don Fernando, of Lord Southesk's breeding, was the field esquire of the milch cows, who do the broom business. The good, solid homestead occupies a square within a hundred yards of the house, and the picked beasts for the great Christmas market swell and fill the stall ranges on two sides of it. The crunch and the groan are sweet music to the soul of "Tillyfour," as enveloped in his plaid he takes his rounds, and watches the rich rations wheeled in from the canteen. How he does hate to see the dust collect on their backs, and what arguments of non-thriftiness he gathers therefrom! They are "their own turnip slicers," as he holds that half the sap is wasted by the more modern system. The caking, except for the more backward ones, does not begin till within six weeks of the great market, when they get 4lbs. to 6lbs. each; but when cake reached £11 a ton, they were principally fed on bruised oats and barley. Peas and beans are no part of their fare.

In contradistinction to the Mechian and Norfolk theories, Mr. M'Combie holds that, as a rule, 14lbs. of cake a day is as much as any beast's stomach can do proper justice to. Only two of the "doddies" had "scurs;" but they were good enough to confirm the butcher's axiom, "never a bad one with a hanging head;" and yet there was only one out of the forty-three which Mr. M'Combie had the smallest notion of "training." He is very particular about a fine forehead and light bone, and if he can "get the tail as small as a rat's" they are always quicker feeders. Fifty-nine were away on the Dee side, but we conned the weekly bulletin of them, and wished that half the civil service candidates could send in as smart a *précis* of the week's doings. Two men were in special charge of them, and the brush and currycomb are not allowed to grow cool in the intervals between the morning and afternoon meals. Only three yearling bullocks were in the house, one of them looking nearly 7 cwt. already, at nineteen months, and another lacking the scale, but very similar in shape to the Poissy ox.

We found the cows with the heifer calves (which are all setoned and oilcaked when weaned) in the pasture close by the house, busy among the new grass left over by the bullocks, to whom they always play second. Foremost among them was the square-made Lovely (by Rob Roy from The Belle), the first heifer at Battersea, and a cup winner at Aberdeen during the time when Mr. M'Combie held that trophy for three years in succession. She still retains much of the style which pulled her through on that day; and Elf of Aberdeen by Black Prince was at her side. The Balwhyllo heifer rejoiced in her Jet of Aberdeen. There, too, was the once well-named Beauty, from Mr. Watson's of Keillor, a fine-sized cow, which fetched 62 gs. at his sale, and she too could boast of a rare Jilt of Aberdeen. This species of nomenclature reached its climax in the calf of Zarah, the second Battersea heifer.

"None half so fragrant, half so fair,
As Kate of Aberdeen,"

ays the old song; and Mr. M'Combie took the hint and named her calf accordingly, and found himself fully justified at Newcastle and Stirling. The dam, which is all going to milk, and has quite sunk her show shape in the matron, was put in price at 90 gs. to a gentleman at Battersea, but he chose three others at £35 a-head, and Mr. M'Combie has his consolation. Zarah was bred by Mr. Collie of Ardgay, and so was Nourmahal, "the dusty haired cow," and the biggest of the bunch. It is very rarely indeed that any owner can say that he won two first and two second prizes in two classes; but Mr. M'Combie has done more, as every one of his Battersea winners has had a live calf, to wit, three heifers and one bull.

In another meadow, Pride of Aberdeen by Hanton formed one of five first-prize Highland Society's winners, which showed side by side at Stirling for their gold medium medals. She is better behind the shoulder, but in her thighs and on the top of the tail she is inferior to her dam old Charlotte. Still youth would be served, when Mr. Hugh Watson late of Keillor and Mr. Graham of The Shawe, two of the finest judges out, of Angus and Galloway stock, judged the pair at Battersea. Charlotte is rising fourteen, and still lacks a whole majority to rival Keillor Grannie. So far she promises well, as there is no patchiness about her, and scarcely any other symptoms of age. Few have been more tried, as she has had foot-and-mouth twice, and lung disease once. Added to this, she had a calf at two years old, and has never missed a year since, but they have generally been bulls, one of which, Defiance by Rob Roy, was sold to the Drummen herd, which has had much local success during the last two years with its females. She began her long list of winnings as first at Inverness, and the first prize at Paris (of which she still bears the brand on her neck) was the result of her only sea journey. She and Hanton were priceless; and therefore the Emperor gave £165 for another cow and £110 for a two-year-old heifer. Walker's Mayflower, which was purchased by Mr. M'Combie for 60 gs., was also here; and Crinoline, another first as a two-year-old at Inverness, but never after, seemed to be wearing better than the Fair Maid of Perth by Angus.

Stepping within doors, we found the walls of the dining-room a perfect epitome of French and British triumphs. Mr. Hall Maxwell looked out at us from the post of honour, as he had done in many a homestead, during our wanderings, and beneath him was the great gold medal of France awarded to Mr. M'Combie "*pour l'ensemble de son exposition.*" The academy of Paris, which seems to devote itself, among other things, to the protection of animals "*sans cornes,*" furnishes a written diploma; but if they could have seen the rush when Mr. M'Combie half-opened the door of Black Prince's box, and our very narrow escape from being pounded to a jelly, they might have felt that their Angus bull sympathies are sometimes misplaced. The hornless are quite competent to take care of themselves. The Albert Cup at Poissy was the centre object of a line of six, which deck the sideboard on high days and holidays. Cups are hard to win; but the trouble which Mr. M'Combie had to get seisin of an English one, when it was won, would form quite an edifying chapter on generalship; and serve as a hint to agricultural societies. The medals have a velvet stand of their own, surmounted by a gold snuff-box (the gift of Mons. Dutrone), and twenty-three gold, forty-four silver, and four bronze hang from its dainty tiers. They are, however, only outward types of a far more solid consideration in the shape of nearly £1,700.*

The first Tillyfour prize taker at the Highland Society is there, dating as far back as 1840, in the shape of a dun Aberdeenshire horned ox, which was sold for £70, and has as his touching Smithfield epitaph, "236*st.* of 8*lbs.* and 28½*st. of fat.*" There, too, in the shape of a black ox from Fair Maid of Perth, is the first-prize winner bred at Tillyfour that ever "burst," not "into that silent sea," but the Baker-street Babel in 1859, or the "Beauty's Butterfly" year. It fetched its £70 and weighed 16¼ cwt. The Bloomer has her place with a view of Windsor Castle behind her; and so has Victor (46), taken when he was not in good condition, and Young Charlotte, "who did no good." The late Mr. Maydwell, of the firm of Maydwell and Hoyland, who had by virtue of his seniority the first choice of the Islington market ground (and to whose firm, as well as Mr. Giblett, the Tillyfour beasts are consigned), has no reason to regret his proximity

---

* The sum total for 1864 was a Fifty Guinea Challenge Cup, five gold and two silver medals, and £207.

to such a glorious specimen as the Buccleuch heifer. "Poissy," with his fine large eye and his ears laid back like a blood horse (no proof of ill-temper, but of the contrary in an Angus) is over the sideboard, looking like life, and faced by those "bloody jades" Pride of Aberdeen and Charlotte, both of which have that white on the udder which has always been popular milk mark of the sort.

Both in point of quality and number of prize "commercial beasts," it was not one of Mr. M'Combie's greatest years. He can tie up 300 on the farm at a pinch, and, in fact, he has had as many as 400 (at home and out on turnips) in hand for market at one time; but last year he did not venture on above fifteen score, and a herd of about fifty —half of them cows and in-calf—made up his home ranks. Twenty years ago he only fed twenty. The heifers are generally put to at two years old, and the calves are dropped as early as possible in the year, to suit the Highland Society, which dates from January 1st. It has generally been the Tillyfour practice to have a sale every other year, and the average in '62 was £32 10s. The calves not kept for the bull trade are never cut before they are a month or six weeks old, and suckled, like the heifer calves, for fully five months.

Hanton and some others of the Tillyfour blood went into Morayshire at beef price, and it is from this county that Mr. M'Combie, who buys every beast himself, draws his principal, and, in fact, eight-tenths of his supplies. Forfarshire, Aberdeenshire, and Banffshire are also placed under contribution, but he does not care much for the Caithness and Ross-shire crosses. "Morayshire for sweetness and quality" is a cardinal point of his creed; and he attributes this superiority in a great measure to the quality of their cows, and their county habit of keeping the beasts in the straw-yards. He would readily give £1 to 30s. more for a straw-yard bullock, as he finds them thrive so much better when they are put to grass. Elgin and Forres are his principal markets, once a month, from December to July, but the owners send him word, and the great majority of the beasts are bought at their own yards. Only one year has he missed the great Elgin April market, viz., when he accompanied his bullocks to Poissy, and then Forfarshire stood in the breach with forty. Captain Kennedy, of Stranraer, in Wigtonshire, used to send him a lot of Galloways every year; and it was from him that he got the black steer, which was first at Birmingham and Baker-street in 1860. These curly heroes of the shaggy frontlet, the thick hide, the odd placed eye, and fan-haired ear, are often better in the thigh, but invariably bigger in their timber and more sluggish feeders than the Anguses; still they will pick up their crumbs royally on the poorest hill land; and this prize winner weighed 14 cwt. clean, and realized £55.

All the bullocks are tied up by the middle of September, and begin to go to the markets at the end of October, in lots of from seven to sixty weekly, and the supply is generally out by the end of March. About 30s. is the average of expenses to London by rail or sea, and last year thirty-nine of the best averaged £38, after all expenses were paid, which gave fully £10 a head for nearly eight months' keep. Three-fourths of them go by the steamer from Aberdeen, and with tide and wind in their favour they sometimes arrive nearly as quickly. In fact, Mr. M'Combie prefers even adverse tides and winds to the eternal shunt, or at times the dreary wait for the missing manifest when they do get to the journey's end. Still, he has only lost condition so far, and none of his blacks have gone down, as the hapless blood yearling Fandanguero did after the eleventh concussion in the station-yard at York, and fairly yielded up the ghost. They came up sixty-two strong, four and five off, to the great Christmas market last year. Eight more went to Liverpool, and sixty-eight of them sold at all prices from £52 to £36, and the other two for £34, and year after year we have the same report from the Smithfield salesmen that "no other feeder had so many good ones in the ranks," and that they died, as of yore, true to their lean flesh charter.

CHAPTER IX.
ABERDEEN TO STONEHAVEN.

> "And then he made some long digressions
> 'Bout cases tried at quarter sessions:
> He talked of squire-detested poachers,
> Of shorthorns, hummels, cobs, and coachers;
> Of antlers in his mosses sunk,
> Of butlers that were always drunk;
> Of controversies about marches,
> And of disease among his larches."
>
> THE SCOTSMAN.

The Royal departure from Balmoral—Up the Deeside—Kincardineshire Sheep-feeders—The Portlethen Herd—The last of Fox Maule—Colour Conception—The late Mr. Boswell of Kingcausie—His Highland Society Essay—From Bourtree Bush to Stonehaven—The late Captain Barclay—The Old Days of the Defiance.

ALTHOUGH we had put the mare in commission once more, we did not wander up Deeside beyond railroads any farther than Aboyne. Her Majesty was leaving Balmoral that morning; and Charles Cook, who now keeps the Huntly Arms at Aboyne had got his brother John, and Davy Troup to take the ribbons again for the day. It was quite like old times, seeing them work their four-in-hands with the Royal luggage-breaks and omnibuses into the station-yard. Alick was also there; but attending to the refreshment-rooms is his present sphere of action. The Royal turnspit was the most troublesome parcel to deal with. He wheeled round on a pivot, and made his deliveries like lightning if any one tried to touch him; and there was a council of four tall footmen over him for minutes on the platform before he was snared and hoisted with a jerk into the "dog-case" of the very last carriage.

We did not care to work up towards Ballater and Bræmar after the West Highlanders, as we were going to touch them at three other places. West Highlanders or crosses with them begin to prophesy of themselves when you get beyond Kingcausie, seven miles up the Deeside. Up to that point, the dairy, which is supplied by shorthorn crosses, blacks, and, in short, anything with an udder, has the pastures pretty well to itself. Sir James Burnett, the owner of the second prize Stirling bull "Prince," is true to the Keillor polls at Crathes Castle; and Mr. John Ross, of Park near Crathes has been rewarded for his 101-guinea venture at Kinnellar by first prizes both at Banchory and Aberdeen.

The dairy system comes in again along the coast as far as Portlethen, where we get among the twenty parishes of Kincardineshire. Fifty years ago, the greatest portion of Portlethen was all whin and moor, and fed no sheep whatever; but the reign of turnips has gradually extended, and hundreds of acres are now let for hogging black-faces off the Grampians. The Messrs. Welsh, father and son, of Inchbreck and Tillitoghills, and Gibb, of Bridge of Dye, are by far the largest sheep breeders and feeders in the county, and have, it is said, flocks of about 1,000 to 1,300 half-bred, black-faced, and Cheviot ewes, which they cross with Leicester tups. They also feed off two-year-old wedders and black-faced ewes for the butcher to a still larger extent, and winter an immense amount of hoggs, of which three-fourths are half-bred, and a good proportion of the rest Lanark lambs. Mr. Gibb, we believe, does not feed two-year-olds to the same extent as the Messrs. Welsh, who are also feeders of cross-bred cattle.

Mr. Walker is the representative of the seventh generation of that name in Portlethen; near whose fishing village an ancient sea captain once erected a steading, and called it "England." His farm marches inland with Kingcausie, and his Angus herd — the only one in Kincardineshire, save those of Sir J. S. Forbes at Fettercairn, Sir Thomas Gladstone's of Fasque, and Mr. Farrel's of Davo — graze close to the coast. About forty cows and heifers compose it, and, with the exception of a few females and a bull to be going on with, the young stock are always sold off as calves or yearlings. "Portlethen" keeps his own vineyard, and has looked out for the best crosses to begin with, and then rung the changes on his own tribes. We were just in time to see the last of Fox Maule, by universal consent the best Angus bull that has been seen in Scotland for many a long year. He was by Lord Panmure's Marquis (212) from Bowie's Matilda Fox by Cupbearer (59), a dam which never failed. Mr.

Martin had been there the day before, and declares that he never killed a heavier beast save one, as he proved at 13½ cwt, *plus* 13 imperial stones of tallow. It was a rare treat to see him come out, with every point so beautifully fitted into each other and bevilled off, and that "neat Roman head set on like a button;" but he was nearly five years old, and had been sadly too chary of his duplicates; and therefore the second-prize two-year-old bull at Stirling, the bloodlike Jehu by Duke of Wellington (219) from Young Jean (295) by Captain of Ardovie (63), was now captain in his stead.

Mr. Walker seldom exhibits except at the Highland Society, Aberdeen, and Kincardine, and gets his full share of prizes, especially with bulls; while "Tillyfour's" strength lies very decidedly in his females. Dinah, Alice Maude, Beauty, &c., are the principal lady-patronesses of the herd, throughout which there are many traces of the old Pityot tribe, in the white below the belly and on the inside and sometimes the outside of the hock. Of white, which is not congenital, Mr. Walker has a very natural detestation. He has had two odd proofs of colour conception—one in a bullock with four white legs and white on the breast, belly, and forehand, which were the precise marks of a strange cat which came about the place; and the other in the calf of a heifer which was chased about the field by a medical man's Newfoundland very shortly after service. The dog was sent for, to compare with the calf, and the white on the near fore foot, hind legs, and tail corresponded with photographic accuracy.

The herd was commenced by its present owner in 1826 with Brown Mouth and Nackets, which were left him by his father. All the Brown Mouths had a brown muzzle, a yellow udder, yellow skin inside their ears, and sometimes yellow stripes down the back, and were not only good feeders, but great butter cows. The blackest-coated tribes will sometimes have a yellow skin, and it almost invariably denotes "a fill pail." This latter quality was very marked in the Nackets tribe, which were darker in their coat and smaller than the Brown Mouths.

Porty by Colonel (145), from the tribe of Rosie, "a dowry cow," whose milking sort had been in the family since '82, crossed well with both these tribes. Colonel was a Nackets bull, with rather a brown back, and so crusty that he had three years of penal servitude in the plough. There was no Aberdeen show in Porty's day; but although he was rather small, his nice shape and peculiarly-fine bone brought him up first at Inverury, and a cross with his own sister helped not a little to improve the quality of the herd. He worked on till he was nine; and the next purchase, after a bull from Mr. Hector of Fernieflat, was Andrew (8), from Mr. Fullarton of Ardestic. Young Andrew (9), from one of the Brown Mouth tribe, was his best son; and then Banks of Dee (12), a purchase from the late Sir Thomas Burnett of Crathes, gave the herd a strong *prestige* in the show-yards with seven firsts and a second. Marquis (212) was bred by Mr. Hugh Watson of Keillor, and was got by his Old Jock, who figures as (1) of the 336 bulls in Mr. Ravenscroft's very valuable Polled Herd Book. It was with Marquis and Raglan (208) by Young Andrew that "Portlethen" stood second and third to M'Combie's Hanton at Paris; and he valued the blood of Raglan so highly, in consequence of his dam Young Miss Alexander (who died from inflammation of the brain through the scratch of a thorn) having only left one other calf behind, that he declined the Imperial offer of £230 and priced him at four hundred. When the bull came back he rewarded his owner for such confidence with Wallace (211) and the Duke of Wellington (219), the best bull one year in the yard at Aberdeen, and from an Ardestic cow.

Mr. Walker does not go very much into sheep, but keeps about fifty Leicester ewes, principally of English blood, round his house, and sells tups and draft ewes. He is also a great Dorking fancier, and had his breed originally from Gordon Castle, where great attention was once paid to it. With the exception of getting a few hens from Ury, and a first-prize cock and hen from Birmingham, he has kept almost entirely to the mealy light hackle of the old Gordon sort. No pains have been spared, but he has had his crosses in the pursuit. Of an expensive setting from Staffordshire only six came out, and of these the foxes, one of which Mr. Fortescue brought to hand in the open last winter with three couple of his Orkney harriers, chose to appropriate four for their share.

Inside the house hang paintings or photographs of Wallace, Banks of Dee, Fox Maule, and Matilda Fox with Fox Maule at her teat. There, too, is Rory O'More, a first-prize winner at the Highland Society in 1847-48. At one time Mr. Walker had sixteen rather small, short-legged, and active workhorses on the farm, all by the gallant grey, and all of them after his colour. It is only three years since the old horse died, and he kept his beautiful shapes when he was upwards of twenty.

One was all for shorthorns and the other for hummels; but in their admiration of Rory and his greys Mr. Walker and the late Mr. Boswell of Kingcausie were one. The portrait of the latter hangs up at Portlethen, "in remembrance," as the donor wrote below it, "of many acts of friendship and good neighbourhood received from an old fellow-agriculturist." Mr. Boswell was the highest example of an improving proprietor. He did not rely on mere length of purse, but he did every thing at the cheapest and most substantial rate, and hence tenant-farmers regarded him as a really safe pioneer. As the youngest ensign in the Coldstream Guards, he had carried the colours at Talavera, and he was wont to tell how he bore quite a charmed life when two colour-sergeants fell by his side, and the flagstick shook in his hand under the hail shower of bullets. After his retirement from the army, he married, and spent fifty years of his quiet, blameless life between his two estates of Kingcausie and Balmuto. One writes who knew him well: "In a careless time he was not ashamed of his religion, and not a few have borne testimony how in earlier years he helped to wean them from the follies of the world, and led them to better things."

Only 250 out of 1,800 acres of Kingcausie were arable when he came into residence; but "the barren, barren muir" had to yield, and hardly a hundred remained not under plough or plantation, when his hand was stayed, and he went to his rest. A lofty Greek cross, which is a well-known beacon to the mariner between Aberdeen and Stonehaven, has been built to his memory by his widow, on the very boundary line where the corn and turnips steal coyly up to the edge of the waste. Of a truth, Boswell's staff stands where he fell. His nephew, Mr. Archer Fortescue, now farms Kingcausie; but the heather acres have not been encroached on, and this last remnant has its especial use as a change of pasture for the hoggs, and a run of a few hours on it during the day while they are folded on turnips has a capital anti-foot-rot effect.

The Highland Society gave him a medal for his success in reclaiming waste land, and never did man set about his work more thoroughly. Some of the deep bog and moss he drained at six to nine feet, then filled it with stones to within three feet of the surface, and top-dressed the moss with the clay. The lighter draining was done on the Deanston principle, thirty inches deep and eighteen feet apart, with stones at the bottom, topped with such only as would go through a three-inch ring; and when some of the drains were opened a quarter of a century after, they were running as clear as ever. The trenching he let off at £10 to £14 per acre, and used the stones for dyking. In carrying it out, he never adopted any paring or burning, but always inverted the top sod and put it in at the bottom, and then ploughed up the field with four horses, fourteen inches deep. The first crop was oats manured with Aberdeen dung, which he bought for two shillings a load at the police auctions, and carted the seven miles during the summer.

Hay he thought quite as exhaustive as an oat crop, and wheat did not suit the land. Wall-building was one of his greatest delights, and he introduced Fifeshire workmen for the purpose. The walls were from four to four-and-a-half feet, with granite copings from his own quarry below The Monument, and joined by cement. They once overlapped about two inches, but this gave the cattle too much leverage for displacing them, and latterly half that projection was all he allowed. This work was, in accordance with his usual rule, paid for by the day, and done by an overseer with picked men under him. There was no premium on idleness, as after half-past five in the morning, wet or fine, they were pretty sure to hear the smart canter of the grey cob in the distance.

He had no eye to planting for game; but "*That's death!*" was his keeper's infallible ejaculation, when he heard the crack of his rifle among the roe-deer. Plantations were an endless source of delight to him. He marked every tree for thinning in his farm rounds, and, when the humour took him, he worked lustily in his shirt-sleeves with saw and axe. Larches had only a troubled time of it, as, although they were sound on the Dee side, they were all piped on the east; and the gravel-pan did not suit them like the black peaty loam upon the yellow clay. Silver firs were his delight, but the bug injured them as well as the oak; while the variegated holly, with its rich, red Christmas berry, "always stood my friend."

Breeding of live stock was a very favourite pursuit with him, and he wrote the prize essay on it in the Highland Society's Journal for 1829, under the motto of

"*Te quoque, magna Pales, et te memorande canemus,
Pastor ab Amphryso.*"

It was upon the comparative influence of the male and female, and enforced the doctrine that to the male we must look for improvement. He ranks Stirling of Keir, Robertson of Ladykirk, and Rennie of Phantassie as his leading Scottish authorities on the point. His judgment on his countrymen was sufficiently caustic, and certainly the comparison between them and the great mass of English breeders does not hold good now. " A great breeder in England," he says, "is a great judge, and one who delights in improved breeds. In Scotland, commonly speaking, it means a man who has a great number of half-starved calves and miserable lambs."

When he first settled in the North, the beasts were "knock-knee'd behind" and narrow. Soon after that, Sir Andrew Ramsay brought a few Lancashire cattle to Scotland, white on the back, with wide spreading horns; and then the doddies came in, "but still the calves cry back." With regard to the horse patriarchs, he "cried back himself" to the days when every grey was a Delpini or a Sir Harry Dinsdale, and every black a Sorcerer or a Thunderbolt. According to him, length of leg was a fault among the earlier sires of the district: Bethlem Gaber to wit, who belonged to Lord Aboyne, and "had the longest I ever saw"; and Buchan's Blaize, near Crieff, who was an immense winner in spite of them. The Suffolk Punch which Captain Barclay brought down he passes over without much notice. Putting Bakewell rams to Highland ewes which had been bred in and in, or " owre sib," till many of them seemed but a handful for a crown, was one of the earliest crosses he noted. Its effects were seen in lambs of the first cross making ten shillings and sixpence, and the wiping out of every trace of the degenerate dam " except a shrivelled horn, which was rubbed off in winter."

By way of testing whether it was more profitable to tie up or feed in open hammels, he selected, in '34, eight two-year-olds and eight threes, on which to try the different systems at Kingcausie and Balmuto. Oat straw and yellow turnips were their fare, and the profit at the end of six months was very decidedly in favour of the younger beasts in the open hammels. He began shorthorns at the estate which came to him in gradual descent "from my trusty cousin of Balmuto"; but, after all, cattle were with him an acquired taste, and his heart was with the greys. When he went to see his friends, Bates and the Maynards, and arrived back with a shorthorn or two, the men at the Aberdeen pier would have wondered what had come over " Kingcausie" if there had not been a mare to unship as well. Two long Kirklevington days with " Tommy Bates," who poured out his whole soul to him in the pastures and over the Duke of Northumberland in the calf-house, were among his happiest in England. His first stock were dun and black horned Aberdeens; then he kept a few polls, and crossed them with a shorthorn bull; and finally, he deserted to shorthorns altogether, and always kept roans, if possible.

At the Highland Society, he only showed fat stock; and he won in the younger class with a red bullock at Aberdeen in 1834. Lord Kintore's celebrated black headed the old class ; and to insure him arriving fresh, he sent his own to Aberdeen a day sooner, and or-

o

dered his men to take on the van with the four grey mares to Keith Hall for "the county champion." "It isn't more than once in a life-time," said he, "that a lord or a farmer has a really crack beast, and every proper respect should be paid it by the neighbours."

Chinese and Berkshire pigs were the "bacon makers" he rejoiced in, on account of their aptitude to fatten; but he was not very particular about breed, and, if the shape was just to his mind, he would buy a pig out of any litter. His sheep fancy was to have fifty black ewes, which he picked up where he could, but he never kept a black ram. At first, he got twopence or threepence more per pound for his wool, which was bought to mix for stockings. He also crossed Anglo-Merino rams with Leicesters, and drafted all those which did not hit to the Leicester form. The wool had a much more silky lustre, but it was his deliberate conviction that no strictly eclectic appetite would have cared for a slice of his mutton.

Grey Stanmore was his cart sire, and begot the renowned Rory from a Clydesdale mare of Mr. Walker's; and he kept a grey Arab to cross his pony mares. He had also a blood horse, Gouty; and this chesnut's daughter, Bessy, may be traced through Myrrha, Ellen Middleton, and Wild Dayrell, down to Wild Agnes. Matching horses in his break was a great pleasure to him; and he pursued the same plan with his Shetland ponies, and sold the pairs at a capital price. His last speculation was an Orkney garron, whose price he nearly doubled when he had kept it a year. It was, however, this love of horses which hastened his end, as a young one ran him against a house and injured his knee-pan; and from that time the fine, hale form which had seen more than seventy summers began gradually to decline. His Shorthorns and Leicester-Merino sheep were sold some time before his death at Bourtree Bush; and people still remember how he rode into the ring, and how, after reminding them of the meaning of a " Bourtree gun," he added that neither it nor a " white bonnet" would be at work that day. After this sale he merely continued the working part of some of his latest improved farms to the extent of about 600 acres of arable, Mr. Fortescue renting the turnips and grass at the average rate of the county, and finishing off on them stock of all kinds which he had bred or bought in the Orkneys.*

"The Bush," where the Defiance changed, was rife with old coaching recollections to Mr. Boswell. He never horsed either the Defiance or the mail, but he would often drive the former for a stage or two. On one occasion, while he and Captain Barclay, who was a great Shorthorn ally of his, were discussing some moot point on the box, the pace insensibly fell off. In about half a minute, Davy

---

* At p. 156, instead of "any disease," in reference to the Orkneys," read "any epidemic disease."

Troup's voice was heard from behind—"*Mr. Boswell, Mr. Boswell! ye'll soon be at sax miles an hoor, and that winna dee, avar*"; and so "Kingcausie" touched up the horses, and the Captain retorted for about the hundredth time by telling Davy to "*touch up your lingo.*"

Mr. Dyce Nichol, of Ballogie, has taken up the Boswell system, and drained and clayed the moss on his estates adjoining the mail road which runs past The Bush towards the South. On the hill to the right is the white castle of Muchalls, looking down on old pasture, which has hardly an equal, save in some of the Home Parks near the Bridge of Dun. There is also rich grass and turnip land at Cowie and Megray; but we only cared just then to know the whereabouts of Ury, which is on the higher road, and touches the east end of the Forest of Cowie, about a mile or so to the right as you enter Stonehaven.

The estate of Ury, &c., on part of which the new town of Stonehaven is built, contains about 4,000 acres, of which the Captain had 700 in his own hand; but all has passed away from "Barclay Allardyce," as he was wont to sign himself, by purchase to the Baird family. His father was a man of vigorous will and industry; and as a proof of it, he thoroughly improved 200 acres, reclaimed 200 more from heather, and planted 1,200 in the space of twenty years. In his speech at the public dinner which was given to the Captain A.D. 1838, in the Glen Ury Distillery at

Stonehaven, he spoke much of the old man, and termed him a "heaven-born improver." The phrase did not apply to his grandfather, who was quite displeased with his son for carrying a bundle of trees on his back the fifteen miles from Aberdeen, and planting them in The Den of Ury. Protecting the plants, he said, would annoy people's cattle and sheep, and he wouldn't have them annoyed. This was just a year before his death, and allowance was to be made for the peevishness of age. When the Captain's father succeeded, he went to Norfolk, where he had first graduated in agriculture, to look out for some good ploughmen. The bosom of the earth at Ury knew peace no more; and turnips and artificial grasses appeared in due season.

There were at least 2,000 acres of baulks, bogs, and rigs on the property, intersected with cairns of stones and muirland, and " the lairds were more inclined to break each other's heads than the treasures of the earth." A smile ran through the company when the Captain gave this out in his deep, solemn tones. They remembered how many men he had backed and trained in his time, and thought him a fine, philosophic compound of the 'two characters. It was a great festival, and nearly two hundred sat down on that July afternoon; but the report of it, which the Captain kept framed and glazed, tells only too sadly that many a "Flower of the Forest" has been "wed awa" since then. Another part of the Captain's speech related to the half-offer which had been made him three years before, to be the governor of a colony of 300,000 square miles in New Holland. The company declared for Barclay, but the Government of the day were against him; and all he knew was, that "some invisible hand checkmated the whole concern." There was a good deal of humorous speculation as to the Captain forming a cabinet. Deacon Williamson and Wetherell would have been very high in office; portfolios must have been offered to the leading members of his Tommiad—Cribb, Spring, and Holtby—Davy Troup would have been Master of the Horse; and of course his friend Kinnear, Attorney-General.

His breeding at Ury began in 1822, with the infallible "*shorthorn* not *shorthorned*," as he always explained, and good-sized Leicesters. Then he bought in West Highland and country cows, and crossed them with shorthorn bulls. For a long time he would sell no heifers, but he soon began regular bull sales. At first £8, and then £16, were thought great prices; when he got to £30 it was monstrous, and as for £60, it was akin to a miracle. He was a very high feeder, and Mr. Wetherell was wont to tell him that others would have kept twice the amount of stock he did on the same grass, and that his cows were very often far too fat at calving; but on this point he was incorrigible. He bought bulls from Earl Spencer, and Mason of Chilton, but none of them were equal to The Pacha and Mahomed of his own breeding; and there is hardly a shorthorn breeder north of the Frith of Forth who does not acknowledge himself under some obligation to one of the two.

At his sale, which took place on Sept. 7, exactly two months after the Stonehaven dinner, Hugh Watson of Keillor bought the first lot, or the "*No*. 20, *Lady Sarah*, 150 *gs*.," of Mason's sale. She was thirteen years old, and her price sunk to 40 gs., but her granddaughter Lily went for 130 gs. to Mr. Allan Pollok, and seven of her tribe averaged 78 gs. This sale was a good one, and of course he took to "shorthorns not shorthorned" again, and kept them till a year or so before his death, when he gave them up, and merely fed some cross-bred beasts. He lacked taste in cattle, and was only an ordinary judge of a beast or sheep; but he had heaps of good sense, and docility to boot, and generally took his cue like a man from masters in the science. In Scotland, Hugh Watson and Deacon Williamson were very trusty counsellors in stock matters, and Wetherell and Jonas Webb stood high with him over the Border. He was not a ram breeder, and when he rather tired of Leicesters he took to Cotswolds, and began to breed in-and-in too much. Cheviots he did not care for, and he had a small flock of Southdowns which he crossed with the Leicester, and occasionally with the blackfaced, and sold the lambs in summer.

When he walked his thousand miles in a thousand hours' match in 1809, his man Cross went at him with a stick to keep him awake, and of course got dreadfully growled at. At that time, he was aide-de-camp to Alexander, Duke of Gordon; and after two days' rest, he went on the Walcheren expedition. He loved to be talked of, and nothing delighted him so much as when he met a regiment on march during his thousand hours, and the officer made them halt, and form in double line, so as to let him pass through them with all the honours. As was natural enough, he was very jealous of his hard-won fame, and never believed in any one, except himself, accomplishing the thousand miles, or in one-half of the "fair heel and toe" feats which he read of. When he was long past sixty, he thought nothing of sending a man on with his dress things and walking the twenty-six miles from Ury to a friend's, and back the next morning. His quiet thoughts on the road were generally believed to hover between shorthorns and getting men into condition. Everything he had to do with, down to his glass tumblers, was always on a gigantic scale. His cattle must be up to their knees in grass, and his wheat waggons—with four or six horses, and the drag on—seemed like an earthquake to the Aberdonians, when they rumbled down Marischal-street to the harbour. Well might the surveyor tremble by reason of them for the safety of the Old Bridge. His bull Champion was cut up for "refreshments" at one sale; and when he thought there might be some mistake about the arrival of the regular beef supplies from the Deacon next day, he had twelve geese killed and spitted on an ashet before the fire. He would have his rounds of beef of a certain circumference, and it was because he despaired of finding a bullock of the regulation size that he made Champion stand proxy.

The same spirit was seen in his management of the Defiance. He would have first-class pace, and he got it. The Union was "the old original," which first went from Edinburgh through Cupar Fife, Dundee, and Forfar to Aberdeen in one day; whereas, prior to its establishment in '26, the passengers spent the night at Perth. Mr. Croall of Edinburgh, and Mr. M'Nab of Cupar Fife, were the principal proprietors of it; and when the Defiance began, on July 1st, 1829, the two often met at Forfar and raced furiously the remainder of the road. The Union was not got up in the same style as the Defiance, as every proprietor had his own harness, and there were no guards in livery. Barclay of Ury, James Scott of Edinburgh, Hugh Watson of Keillor, Donald Seaton of Bridge of Earn, and Captain Skelton of Kinross, were the first proprietors of the Defiance, which took the high road by Forfar, Cupar Angus, and Perth to Queen's Ferry. Davy Troup, a *protégé* of Captain Barclay's, was one of the earliest coachmen; and Henry Lindsay was such an active guard, that he would drop off behind, run round the coach, and on again when it was going its best pace. Twelve miles an hour, including stoppages, was the regulation speed; and it kept time so exactly, that half the watches in Stonehaven (the scene of the ten-minute breakfast) were kept by it. The horses were generally three-parts bred, and worth from £30 to £40; and Mr. Watson, who horsed the thirteen miles from Cupar Angus to Perth in connection with the late Lord Panmure, had a couple of sixteen-hand Yorkshire mules as leaders in one of his stages. More confirmed "merry-legs" were not to be found in the whole 126 miles. Captain Barclay horsed three stages from Stonehaven to Northwater Bridge; and each guard undertook one. Only a minute was allowed for a shift; and the pace was so steady, that when the heir to a peerage, who was going to fight the Elgin burghs, fell asleep at Perth, and missed the coach, he never could sight it again in the eighty-four miles to Aberdeen, although he was turned out very little more than ten minutes behind it with four good posters. Seven o'clock was the time of starting from Aberdeen, but the Queen's Ferry cut the time to waste, and Edinburgh was not reached before eight.

The first coaches had blue bodies, and red wheels picked out with straw, and cost £150 each in London. They were beautifully hung, and so low that they could not be overset; but the draught was too great, and red ones from Wallace of Perth gradually superseded them. In 1838, the coaches were stopped a whole week by snow, at the time when a couple of wedding-parties as well as Ducrow and his *troupe* were blocked up at a public-house near Bervie; and spotted horses and trick ponies were distributed with fairies, acrobats, and the like, among all the adjacent farms. Nine or ten mails were due in Aberdeen at that date, and the Cooks all but perished one night in riding on with the bags. Mr. Watson gave up the Defiance in '36, but, with the exception of a very short interval, Captain Barclay's connection with it was unbroken till the end. Then Mr. Seaton died, and his widow (who kept on the Salutation, at Perth) and Mr. Elgin took the ground between Perth and Cupar Angus. The latter had a stage of the Glasgow Defiance along with Mr. Ramsay, who came into the old concern; and as Mr. Croall had given up the Union and joined hands as well, matters at the end of the first week of 1842 were remarkably flourishing. The two Defiances now used the same livery and account-books; and Mr. Croall furnished coaches with patent drags at so much per mile. In fact, there was only this solitary distinction left, that the Edinburgh coach retained its brass-mounted, brown leather harness, and the Glasgow its silver-plated, black leather. The Defiance never ran on a Sunday, and its take for the six weeks ending August 26th, 1843, was £2,216 3s. 8d. Of this, £1,422 10s. 7d. was divided among the horse contractors, who paid their own strappers, and the remainder went in tolls and mileage, &c. The tolls alone for that period from Queen's Ferry to the Bridge of Dee were £135.

Mr. Ramsay was then only four-and-thirty, and in

the very zenith of his coaching enthusiasm. Sheriff and Fulton were his saddlers; and if he ordered or did a thing to-day, it was the fashion all over Edinburgh to-morrow. He had started the Quicksilver against Croall to Newcastle, and the Tallyho from Edinburgh to Stirling; he had bought up Steele's business on the Hawick road for something like £4,000; he had built the May-day yard at Stirling, and stationed quite a colony of horses there for the Glasgow Defiance and the Rapid, which also ran from Stirling to Perth, but round by Crieff. At one time he had four pairs of beautiful greys at Forfar, and he would occasionally work the Defiance the whole journey when he was in the humour. The three Cooks, John, Charles, and Alick, came over to the Defiance with Mr. Croall, and Charles, and Davy Troup worked the coach from Cupar Angus to Aberdeen, while James Lambert and (on his death) John Lowden, another Union man, and little George Price, with his natty, blue bird's-eye and pin, took the rest of the journey. Lambert had a great knack with his whip, and if he passed any pigeons or chickens on the road, he would turn round and ask a passenger which he would prefer for lunch, and as quick as thought whip it up on to the coach with his thong. The liveries were red coats with yellow vests, white hats, and silver-plated *" Defiance—Aberdeen and Edinburgh"* buttons; and all the guards had patent time-pieces.

Neither John nor Alick Cook was such a musician as Tom Godwin, who was on the Glasgow Defiance.

Crowds used to gather round the coach in Perth to hear him give "*The girl I left behind me,*" and "*The days when we went gipsying;*" and he threw such intense feeling into "Rory O'More," or, as his audience observed, "*played it with all the gestures,*" that, long after he became a landlord at Stirling, he went by no other name. Alick Cook was a smart little light-weight, and went about the coach like a needle. He was also a great cock-fighter, and at every thing in the ring. The Shadow ran several races in his name; and he was generally looked upon as a sort of *Racing Calendar* guard; well up in the likely starters for the St. Leger, and more especially in the moves of the green and yellow jacket of Barnton.

The birthday of the Defiance was kept with great state at one of the towns on the road, and the proprietors and their friends feasted out of the receipts. On one occasion the coach was dressed up with evergreens, and the horses in flowers and streamers. "*Alick Cook, music for anniversary, £1,*" was always an item on the debit side of the accounts for that month. The Captain invariably drove the coach on the dinner-day in a scarlet coat, and at night he was ready with his "Trotaway" song respecting the mare, who was on her legs like the shot of a pistol, and beat the bullet cleverly when she was there. The singer nearly lost his life after one of those festive evenings. He had not become habituated to gas in his bedroom, and on retiring to rest at Forfar he merely blew out the jet. Awaking nearly suffocated, he at once relieved himself in his crude, emphatic fashion, not by groping for the door, but by delivering "one, two" straight from the shoulder, through the window-panes. He was not an elegant, but a very powerful man on the box, and even during his proprietorship he always fee'd both guard and coachman whenever he went with them. Davy kept the interests of the coach most rigidly in view, and spared no man if they were jeopardized. When the Captain slackened his pace in the midst of his conversation, he used no circumlocution, but exhorted him from behind to "*Gie us mair of your fup and less of your claiver;*" and when the Captain just grazed a cart, he was at hand in his Aberdonian Doric to caution and correct: "*Fat's the use, Captain, of takking an inch on the ae side, fan there's ells on tither?*" A little Perth colloquy was overheard between them when the Captain had just finished his dinner at the Salutation. Davy was taking occasion to observe publicly that the Captain had "*gat a braw red face;*" and he was receiving an elaborate explanation to the effect that it was entirely owing to the good dinner, and the glass of punch after it.

The Captain was unable to attend to the inner man so narrowly, when he had trained Sandy M'Kay, and was taking him up to the fight. People soon learnt, who that ugly, slouching, round-shouldered man was, on the box-seat, dressed in a cutaway and top-boots, and why the Captain never kept his eye off him during dinner, lest he should exceed his three chops and a glass of ale. The poor fellow had a gloomy foreboding that he was being driven to his doom; and when it proved too true, the Captain had to get into a Dundee smack, and then hide for a time in Forfarshire till the Government tired and dropped the prosecution. Five years before his death, the horn he loved so much ceased to be heard in Stonehaven; and in the last week of October, 1849, when the route was reduced from Aberdeen to Montrose, the Defiance ran its last journey, and bowed its proud head to steam.

The Captain was, as one of his most intimate friends summed him up, "a great eater, of fine, simple faith, and always in condition." When he first met Hugh Watson it was at a coursing meeting, and seeing that he was a man after his own heart, he asked him, as if it was a highly intellectual treat, "*Would you like to see me strip to-night, and feel my muscle?*" He denied no "Pug" in distress, and when he gave his celebrated supper to the "Fancy" at Tom Spring's, there was such a gathering of the clans that the police looked in, and two or three guests were "wanted" in the course of the evening. To this day they speak of him as a departed pillar of the profession. He trained Cribb for his Molyneux engagement on Scultie between Ury and Drumtochty, West from Stonehaven; and the sight of "lofty Moss Paul" impressed him so deeply that he straightway confided to the coachman, "*I should so like to train a man there*" For him scenery had no other significance than as an adjutant to condition. Jackson was often down at Ury, and they visited the late Lord Panmure at Brechin Castle together. This *fidus Achates*, Gully, Cribb, The Game Chicken, and Barclay himself in a cock and pinch hat and a yellow handkerchief, as he appeared when walking the match, were the principal adornments of his dining-room; while little, fighting portraits of minor lights, with shaven heads and broken noses, hung in the porch and hall. Snowball, the greyhound of the Wharram Wolds, and his dog Billy were also held in honour; and formed part of his gallery of heroes, human, pastoral, and canine.

His dress was curious, but still it never concealed the high-bred gentleman of primitive tastes. He had generally a blue or yellow handkerchief round his neck, and a long yellow Cashmere waistcoat. In summer he wore a green coat with velvet collar and big yellow buttons, coarse white worsted stockings, as often as not a patch on his knee, and very wide shoes. The dog-days would bring out a white linen jacket and moleskins, and he had always a little quid of tobacco in his mouth, to which he gave one or two rolls before his long, measured speech began.

He once kept a pack of foxhounds at Allardyce near Bervie, the estate from which he signed himself "Barclay Allardyce," and hunted Kincardineshire and the Turriff countries, sometimes riding forty miles from Ury to a meet. Latterly he went to Leamington for six weeks during the hunting season, and made quite a sensation at one of the balls in his green coat and black knee-breeches. If, as he complained, some of his old friends had forgotten him in his yellow buckskins and venerable mahogany tops, which generally rested pretty nearly on his instep at the cover-side, they knew him fast enough at night. Of the Aberdeen races he was once a great supporter, and ran horses; but all his comrades died off, and, latterly, beyond betting a trifle and having Fancy Girl at I'Anson's, he was almost out of it. He was generally at Epsom, and he used to tell with some glee that the most nervous men he ever saw in a road-lock on the Derby day were two of the old Defiance coachmen.

At home his own habits were very quiet and simple. He was always ready with his subscription for any good object, and every Monday twenty or thirty people would be waiting for him about the front door after breakfast for their sixpences, of which he carried a supply in his waistcoat pocket. On New Year's Day he had always his friends to dinner, and he sat obscured to the chin behind the round of beef which two men brought in on a trencher. Mr. Kinnear was the perpetual Vice, and every body made a speech. The Captain's was quite an oration, or rather a *résumé* of the year, and concluded with a special eulogium on those who "have died since our last anniversary." Not unfrequently he killed one or two before their time, perhaps more from a little dry humour than by mistake; and then he begged their

P

pardon, and said, "*It didn't matter much.*" For some time before his death he had suffered slightly from paralysis, but a kick from a pony produced a crisis, and two days after, when they went to awake him on the May morning of '54, he was found dead in bed. He lies in the cemetery of Ury about a mile from his old home—the trainer of pugilists with the gentle apologist for the Quakers—and his claim to the earldom of Airth and Menteith seemed to die out with him.

CHAPTER X.
STONEHAVEN TO CORTACHY.

"And as for cattle one yearling bull
Was worth all Smithfield market full
Of the golden bulls of Pope Gregory."
HOOD.

The Weathercocks of Stonehaven—The Sea Coast Road and The Mearns—Stock of the District—Bulls at Fernyflatt—Yorkshire Calf Trade—Angus Commentaries—The Kinnaird Valley—Old Montrose—Black Herds—The Kinnaird Castle Steading—Druid and Cupbearer—The Great Forfarshire Covers—Cullow Fair—Cortachy Castle—Highland Crosses—Laying down Permanent Pasture without a Crop—The Cortachy Herd and Dairy.

THE pigs, which, owing to their presumed capacity of scanning the "viewless forms of air," surmount most of the Stonehaven weathercocks, stood at 7 a.m., with their snouts due east. We had had a wild night of it, with the waves at very boisterous play just behind the Ury Arms; and the wind blew bitterly as we scaled the heights of the coast road towards Bervie. Our line of country was through a deep, alluvial loam, while the old Defiance road, which lay some miles to our right, winds its way through the lighter and more "gravelly soil of the Mearns." A large amount of potatoes, beans, and wheat is grown between Bervie and Montrose. The pastures and part of the turnips near the sea are let during the winter for blackfaced hoggs and wedders

P 2

from the range of the Grampians; and this system
begins at Girdleness, and continues along the coast,
and through the Mearns as well. "The little men of
the Mearns" have arable farms of between two and
four hundred acres. They breed very few cattle,
and buy what they want in the Falkirk trysts or North
of Aberdeen, mostly two-year-olds, to finish and turn
off at three. Blackfaced wedders are sometimes bought
and turniped along the coast; but farmers in these
parts, as a general thing, do very little with sheep
on their own account. The Southdowns, which we
had almost lost sight of since Gordon Castle, except
at Mr. Walker's in the Aberdeen district, began to
rear their heads again as we drew nearer to Keillor,
once their great Scottish stronghold. Mr. Garland
of Cairnton breeds and sells tups of the sort, as well
as Leicesters. The latter are used to draft Cheviot
ewes from Sutherlandshire, and at Fetteresso near
Stonehaven the blackfaces have been crossed most
successfully with the Southdown. Two blood sires,
Champ Fleury (so called after a celebrated West-
Lothian meet), and Cortes by Alarm (bred by Mr.
Greville and the sire of Jessie), travel about here,
but it is not much of a horse district, and the youngest
farmers continue faithful to life in a gig.

Mr. Arthur Glennie, who has, in proportion, nearly
as large a practice as judge at shows in Scotland as
"Mr. Baron Unthank" in England, farms Fernyflatt
which lies between the sea and the road. He breeds
a few good polls, and feeds off annually, among

other stock, a large number of bulls. We found no less than twenty, nearly all of them shorthorns and from the North of Scotland, stowed away in every nook and corner of his buildings, and had a faint realization of Bashan at last. There was every inducement to linger in that snug parlour, which was Lincolnshire to the life, but we had a heavy list of calls in our note-book, and a really fine afternoon was not to be spurned. The spray was dashing over the little pier of John's Haven, and every thing along that grand sea board was so full of life, that we seemed to feel for the first time that our heavy task was really beginning to " give."

There is some fine old grass land at Brotherton beyond Bervie, where Mr. Hercules Scott keeps Balwyllo blacks and Sittyton shorthorns. Between Brotherton and Montrose there is very little grazing, but beasts are simply wintered and sold off gradually to the London and Glasgow dealers. Mr. Miller, the Montrose butcher, buys about fifty a week for Glasgow; and Mr. Scott sends a great deal of dead meat as well both to London and Newcastle. It is a peculiarity of the latter market, which may be traced in the biddings at Earl Durham's annual sale, that it does not care for steers nearly so much as neat cutting heifers, both in winter and the hot summer months. Nearer Montrose and down towards Forfarshire the farmers do much more in store beasts, which some of them get from Falkirk; but the Yorkshire calves are still in the highest request. They are of all ages from six months to a year, and cost from £4 to £7. Jackson, who has a park at Guthrie, will bring down five or six hundred, and Luke Land has a good supply as well. Some yearlings, which range from £8 to £11, share the trucks with them, and throughout September up to the middle of October a very brisk trade goes on. Some of the heifers are kept on to breed from; and one farmer in Aberdeenshire lately purchased more than forty for this purpose.

Cross-cow breeding is on the increase; but many farmers seem to use anything with a hide and a stomach; and North of Aberdeen more especially there was too often a sad lack of loins among both pure shorthorn and cross-bred beasts. As for a Water Esk beast, it is frequently a compound of Shorthorn, Poll, and West Highlander. It is great fun starting a regular doddy cynic upon the Yorkshire question; and "rough things, heifers and bulls all grouped together, and not two inches of spine the same level among the lot," is the invariable commentary. "As for colour," says another of these unwavering ebony standard-bearers, "I tell my friends they ought to get a green bull, and then they'll be all round the rainbow."

"*Pass on, Traveller, safe and free of toll,*" is the inscription on Northwater Bridge, which boasts of three architects. After sleeping at Montrose, we passed with due reverence the statue of brave, old Joey Hume in the "canny" heart of his native town, and made our way over the chain bridge towards the valley of Southesk, and down on the heavy clays of Old Montrose by the little ivied church of Maryton. We had not seen such land since we bade farewell to the early districts near Alves. The Howe of Kinnaird, as it is sometimes called, begins at Montrose, and extends six miles by three to Brechin, and joins the Howe of Mearns just about the point where the latter melts away into Strathmore. The whole valley belongs to the Earl of Southesk, and his castle of Kinnaird holds a wooded eminence in the very heart of it. Those pinnacles mark the only poll herd in the valley, and, in fact, there are nothing but Shorthorn crosses nearly all the way to Arbroath.

The Earl of Southesk's factor, Mr. Charles Lyall of Old Montrose, Mr. Goodlet of Bolshan, and Mr. Swan of Inverpeffer, are almost the only sheep-breeders in this district. Mr. Lyall once divided his allegiance between Leicesters and Southdowns, but he has now given up "the copperheads," and retains a flock of Leicester ewes, principally of Cockburn of Sisterpath blood with an infusion of Sanday. Mr. Goodlet has a flock of half-bred ewes, to which he has latterly been using the Kelso tups. Three-fourths bred, half-bred, and grey-faced "mule" lambs are all at high pressure on cut turnips, cake, and grain during the winter. Some black-faced wedders are also busy among the turnips in the

Vale, but they generally belong to dealers. In the "Shire of Angus" tares and turnips have increased of late years to an immense extent, and have quite taken the place of bare fallows. Potatoes flourish all the way from Old Montrose to Perth. Regents command a higher price than Rocks; but the crop of 1863 was so tremendous, amounting in some districts to ten tons the imperial acre, that they had almost to be given away, and no one got more than a pound per ton. Hence the farmer was just as well off during the great disease of '46. Six tons are a fair average crop, and £3 the price of ordinary seasons. The potatoes thrive best on the black-loam edges of the Kinnaird valley, and beans on the strong clay of the flat. The valley was at this date the northern limit of the seven or eight steam ploughs of Scotland, and the very first that crossed the Border, to the order of a private purchaser, came in '61 to Mr. Lyall's farm at Old Montrose.

This farm-steading has a very warm, English look about it, with its sycamores and Spanish chesnuts, and its old garden walls; and it was refreshing to see "a bit of Bates" at last in "Little Go" by Fourth Duke of Oxford, who was purchased from Earl Airlie. The herd began with "Southesk," a Bates bull, one of the only four calves which Second Duke of Northumberland left, when Captain Barclay hired him in '41 for a season. Prince Ernest by The Baron (now Bowley's Little Go) did more in his day, and claims Leonora and the good-ribbed Ruby among the twenty pedigreed cows, of which Rosemary and the broadbacked Mysie 10th, Jessamine from Sittyton, Barber's Duchess of Northumberland, Towneley's Duchess Nanny, and Queen of Beauty (who was too proud to go with the others), all caught our eye in the pasture.

No particular attention is paid to pigs in this district; but Berkshires have been introduced through Mr. Charles Carnegie, M.P., who took a fancy to them when he was a student at Cirencester College. They are thrifty feeders, and hold the yards in winter along with some small whites of Brandsby descent. Dealers go round and buy them up when they are between seven and eight stone, and pack off the carcases in crates to London. Clydesdales come for the season from the Glasgow districts, and Mr. Wilson of Portsoy, has more than once sent some of his champions to take the prize at the Angus Agricultural. This show takes place just before the Highland Society, and includes in its circuit Forfar, Dundee, Montrose, Brechin, Arbroath, and Kirremuir. Horses, like the farms, do not grow larger as they approach the Grampians. The General, who was then in Mr. Charles Lyall's hands, left some good stock about here during his three seasons, and so did Lord Strathmore's Master Robin. The blood of Lord Panmure's Cleveland has worn pretty well; but not a soul would look at The Cure in the days when he was down in the world.

There was once no better polled stock to be found in Angus than at Fullarton's of Mains of Ardovie, whose Queen of Ardovie was the dam of M'Combie's Queen Mother. The Keillor herd is dispersed, but the rest keep Aberdeenshire, Kincardineshire, Banffshire, and Morayshire very fairly at bay in the Highland Society's lists for the honour of the "shire." Besides Lord Southesk and Mr. Bowie, Mr. Mustard of Leuchland is a very old breeder, and has always crossed with Keillor and Kinnaird Castle blood. Lord Southesk's brother, the member for the county, keeps a few at Arrat's Mill, which has always been celebrated for its old breed, and bought some more at the last Balwyllo sale; and Mr. Goodlet of Bolshan posed the two Walkers, who were represented by Jehu and Sambo, with his Dahomey by Windsor, from Oriana by Cupbearer (so called after his noble owner's State office), in the two-year-old bull class at Stirling.

Mr. Lyell of Shiel Hill, near Kirremuir (brother to Sir George Lyell, the great geologist), who dipped well into Southesk Mariner, had a first prize with his Prospero at Battersea, and a third with Commodore Trunnion at Kelso; and Mr. Leslie of The Thorn, Blairgowrie, defeated the Black Prince of Tillyfour and the Black Diamond of Montbletton, both for the first and second yearling prizes, at Stirling, with his President 3rd blood. Mr. Scott of Balwyllo died in 1842, since which time the herd has been carried on by his widow. He was always great for heads and necks, and took prizes with his slashing year-olds, whose size atoned for a slight family sharpness in the hair. The prices at the '63 sale were well up to the mark. "Portlethen" and "Tillyfour" fought for Alice Maud as far as 62 gs., and then the former was up to time with another guinea, and got her, while Heather Bell 60 gs. became Lord Southesk's.

His lordship's steading lies about three miles from Balwyllo, and about half way up that long avenue of beech, elm, and sycamore which leads to the castle, for which nature seems to have reserved all the variations of ground. There are about ten thousand acres in the whole estate, of which 1,500 are within the park fence. The herd has been in the family for fifty years, but the present one may be said to have taken a fresh start since '52, when his lordship came to the title; and Cupbearer by Pat (29), from Rose by The Colonel, has been its great mainstay. Mr. Walker bought him from Mr. Bowie after he had won the first Highland Society prize as a two-year-old at Perth, and got £20 for his bargain. He was a very cheap one to the Earl at £55, as he was used for five or six seasons, after which he wasted so fast that he had to be killed without loss of time. His back was rather slack as he stood, but "he swelled when he moved," and several of his stock inherit his white beneath and on the inside of both thighs. A great number of bulls by him were sold, but the best was kept, to wit,

Druid, from Dora of Ruxton of Farnell's breeding and of a Keillor strain. This cow was a very remarkable breeder, and never missed, except upon the occasion of her Paris trip. She also bred Dulcinea, an own sister to Druid, and a first-prize taker at the Angus show. Her Diodorus by Windsor took a first International prize at Hamburg, and was sold to Russia; and her Kathleen by Strathmore (one of old Grannie's calves) is now quite a herd notable, as she is the dam of Calliope by Raven, a Druid bull, and granddam, through Calliope, of Clio, the first-prize yearling at Kelso.

The Druid won more than his sire, and, including the £25 at Chester (where he and his sister Dulcinea and his half-sister Oriana took firsts in three classes), he gathered up £100 among the show-yards in his day. He was more of a show bull than Cupbearer, more level in his points and longer in his quarters, and with finer style altogether; but although he was used for a season or two more in the herd, he did not leave such a mark for good. He was sent to Battersea when he was eight years old and had quite lost his bloom; and although the judges allowed that he was "perhaps in some respects better," despite his age, they could only place him second to Prospero, "a bull of very fine quality but small in size." The portraits of him and his sire hang side by side in the Castle library.

Windsor succeeded the Druid, and took the first

old-bull honours at Edinburgh in '59, as the Druid had done at Glasgow in '57, and Cupbearer at Berwick three years before that. Of the other purchases, M'Combie's Empress, which was a calf at the foot of Charlotte in Paris, has done well for her 60 gs., and so have Emily by Old Jock, Walker's Princess, (a great milker,) and Balwyllo's Queen, which was second at 7½ years, at Kelso, to Wemyss's Nancy, a Tillyfour cow of no great touch, but well made up, and with youth on her side. The herd scarcely ever competes, except at the Highland Society and Angus Shows, and, fortunate as it has been with bulls and heifers, it has never yet won a Highland Society first with a cow.

The steading, which cost under £4,000 and measures 270 feet by 132, is a very great feature, with its platforms and beautiful slate ventilation, and, above all, that grand sense of airiness and Australian elbow room, which results from its cattle courts being half open and half under cover. Several cattle were put up to feed, and among them was a red Angus with a white nose. Antiquity will have its say, and some poll breeders maintain that no statute of limitations prevents the blacks from throwing yellow, red, and brindled calves at times. Mr. M'Combie has had calves of this colour, but not within the last seven years, when he sternly vanished "every beast of colour" from his breeding farm. Quadroona by Windsor from Queen (445), and quite the young black belle of Kelso was by her side calf-less, and with Smithfield, where she was first in her class, as her future portion; but we had to look afield for Clio by Windsor and Clarissa by Don Rhoderick, which were first and second in the younger class at Kelso, and kept their relative places at Stirling, where a Montbletton heifer beat them both. Clio would be an excuse, if there had been no Quadroona, for liking the Windsor heifers.

There were racks for thirty-six in the byre, where Old Violet stood fallow at last in her sixteenth year. She has been one of "the Kinnaird milkers," up to her twelve quarts at a meal, and in regular descent from Old Bell and Lady Anne. Windsor was looking wistfully out towards the pastures in one of those ten cow and calf boxes with yards, at the south end of the building, which are all furnished with double sliding doors. His fine quarters make you forget his roughish shoulders, and as he stood foreshortened in his yard, with a field of golden barley beyond, and the Castle pinnacles rising through the trees above it, we thought we had never seen a more beautiful picture to hand.

The vast Forfarshire woodlands were wont to be full of foxes, which have suffered in their time quite as much from politics as from hounds. Two-thirds of the three thousand acres in the great common forest of Montreathmont or Mountroman Moor belong to Lord Southesk. In some parts it is fully two miles across, and is principally planted with Scotch fir and larch; but the rough heather bottom is so wet and cold that there is not much good lying. In frosty weather "it was beautiful running when other packs were idle." Tom Rintoul (who was wont to boast that the majority of huntsmen "wouldn't get a pack out of it in eight days") killed 6½ brace of foxes from it in one season, as it is quite a city of refuge for them. It is also full of roe deer; but Tom's hounds had been well broken to "*War! Hench!*" before they came there, and therefore they took no notice of "the finest scent in the world." Legaston is a capital 200-acre gorse, and so thick that, to quote Tom again, "my hounds used to come out with their sterns as bare as a whip-thong, and so bloody that I hardly knew them." Catterthune above Brechin was another fine cover; and so was Marcus, five miles to the west, with a splendid Leicestershire country, and a burning scent over heather right away to the Grampians. Colonel Maule, who died in the Crimea, was the last regular master; and when Major Douglas was appointed field-master, he bought 25 couple of the Hursley for £300, and, with Ben Boothroyd and Markwell as his huntsmen, carried on the game for four seasons more. A better sportsman never spoke to hound; but all his fine riding did not prevent him from leaving one or two of his best horses dead upon the hills above Cortachy, and trudging back with the saddle and bridle on his arm, when a straight-necked, dark-coloured "traveller" had "gone home in a hurry" from Marcus. Mr. Hay, of Letham Grange, is another grand, old county sportsman, and has had harriers for more than forty seasons. Fitchit has long been his trusty aide-de-camp in stable, field, and kennel, and they are not severed on Mr. Gourlay Steell's canvas.

Tom Rintoul always speaks of Kirremuir as "a rough place," from the fact of his having been nearly mobbed one night as he was bringing the hounds through it during a weavers' riot. The shuttle goes flashing through the loom with a sound which seems very foreign to an ear which has been accustomed for weeks to nothing but the bleating of flocks and the lowing of herds; but it is worth climbing up that weary hill towards Cortachy, if only to see the panorama of the Vale of Strathmore. We were too early for Cullow, where the blackfaced wedders—the gatherings of the glens and the Grampians from Glenshee to Glenesk—are mustered seven to ten thousand strong about the 18th of October. Ballater and Bræmar furnish many of the wedders, and if they are not sold at Castleton fair, they come on here as "Grampian sheep." Four-fifths of the whole are really Lanark lambs, which are bigger and better bred than the mountaineers. The predominance of them makes this the best blackfaced wedder fair in Scotland; factors buy them up as twos and threes chiefly to stock the policies; and Mr. Geekie, sen., will occasionally take as many as fifty score for Earl Mansfield's home farm at Scone. A great many go into Forfarshire and Perthshire, and some few into Fife. Last year the top price was 40s., which is the highest on record; and the lucky lot came from the grazing farm of Mrs. Kennedy, at Glenmaye.

The Highland games of the clan Ogilvy were in progress at Clova, but we did not care to wend our way ten miles in search of them, and perhaps get into difficulties behind Caen Lochen, and his cold, rival height of Catlaw, both of which are sometimes powdered with snow in June. A clear frosty night had blackened the potatoes and the fir cones when we left our snug ale-house " hole in the wall," and made our choice for the day between Clova and Cortachy. This "bonnie house of Airlie" is in a very beautiful spot five miles from Kirriemuir, and just under the well-wooded spurs of the Grampians. The first two or three miles are not inviting, but the rustic seats and fountains along the road as you approach the Castle grounds give you quite a "Rest and be Thankful" feeling, and help to cheat the toil.

A small flock of Shetland sheep, black, white, and black-and-white, were in Craigiss field. They had been there nearly a year, but they were still as shy as rabbits, and would smeuse anywhere. Only a few of the lambs have been killed as yet; but the second-year fleeces weighed from $2\frac{1}{2}$ to $3\frac{1}{2}$ lbs., and the ewes bid fair to kill at 10 lbs. a quarter, so that a more generous diet has worked well. The West Highland crosses were in the snug lying of Ball Field on the opposite side of the Southesk. They were by a shorthorn from black cows, and the colours have fallen black, dark brown, black with a brown streak, black-and-white, and grey. They take the form of the sire very much, with the exception of the head, neck, and horn, and look pretty nearly as large when yearlings as a West Highlander at four. Perhaps for this county the cross with a pure Angus bull struggles better with a hard winter. The characteristics pair off precisely the same way, but the colours are almost restricted to black and blue grey.

Grass is not too abundant in a Grampian climate. There is not much to speak of till nearly the middle of June, and it begins to fail early in August. Since Mr. Peter Geekie came, as factor, the Earl has adopted the plan of laying down all his permanent pasture without a crop, and it has so far proved a much more profitable method than the old one. There is no risk of a wet summer laying the crop and rotting out the grass. The sward is compact and firmly rooted before frost sets in, and is consequently not so readily thrown out, while it is better prepared to brave the winter, and come away early in the spring. It stands to reason that it should be better the second year for more causes than one, but chiefly from the fact of its not having been nursed and drawn up in the shelter of a crop during the mild part of the season, and left after the cutting of the crop exposed to the bleakest and coldest weather.

The following mixture of seeds was used by Mr. Geekie in sowing down permanent pasture without a crop :—

Field No. 1 was sown in April, 1863, after turnips. Before sowing, it was carefully drained 4 feet deep and 24 feet apart ; afterwards ploughed and sub-soiled, and well limed with about 4 tons of lime per acre; then ploughed again, and harrowed and rolled till a fine seed-bed was obtained, when the seeds were sown and lightly harrowed and rolled in. By the middle of July it afforded a free bite for sheep, and before winter set in there was a beautiful, close sole of grass. This year it has been most luxuriant.

Nos. 2 and 3 were sown down this spring, and are now excellent pasture. Both were sown after green crop, and the land prepared in the ordinary way as if for sowing down with a crop.

In sowing down in this way, weeds are very apt to get up in large quantities and choke the young plants; and when this occurs, no time should be lost in "skimming" them over with a scythe. The grass gets up very quickly after this operation.

Where the field is exposed, it is a good plan to sow half a bushel of barley along with the seeds; it gets up faster, and affords much shelter to the young plants. In sheltered situations, however, it is better not to do so; for if the season is moist the barley is apt to get up too quick and strong, and smother the grass.

The price of mixture No. 1 is £1 14 8 per acre.
,, ,, No. 2 is 1 7 4 ,,
,, ,, No. 3 is 1 12 7 ,,
but of course these prices will vary with the rise and fall of markets.

Field No. 1.—Soil stiff, sandy clay, with gravelly subsoil; exposure southern.
Field No. 2.—Light sharp soil, with gravelly subsoil; exposure southern.
Field No. 3.—Soil very light open gravel; exposure northern.

Seeds.	Field No. 1.	Field No. 2.	Field No. 3.
	Bushel.	Bushel.	Bushel.
Pacey's perennial ryegrass	½	½	½
Stickney's ditto	½	½	½
Annual ryegrass	¼	¼	¼
Italian ditto	¼	¼	¼
	lbs.	lbs.	lbs.
Cock's-foot	3	3	3
Meadow Fox-tail	1¼	1¼	1¼
Sweet Vernal	¼	¼	¼
Hard Fescue	2	2	2
Meadow ditto	2	2	2
Sheep's ditto	1½	1½	1½
Timothy	1	1	1
Rough-stalked meadow-grass	2	2	2
Smooth ditto	2	2	2
Alsike clover	3	3	3
White clover	4½	6	6
Dutch red clover	3½	3	3
Cow-grass	2	—	—
Welsh red clover	—	2	2
Sheep's Parsley	—	—	4

The church, with its thick yew and laurel girdle, stands hard by the Castle-lodge on the banks of the Southesk. We look through the window of the family burying-place and read on marble how the fifth Earl of Airlie, "in the generous enthusiasm of youth joined the Chevalier at Edinburgh, 1745, with a regiment of six hundred." The inscription on one old, grey tombstone records the frail assurance of a survivor for its inmate,

> "Who by a course of verteous acts
> Eternal life secured;"

and a poet has evidently been engaged to set forth the crowning act of another Forfar worthy, who died in 1732:

> "Here lyes James Winter,
> Who died in Peathaugh,
> Who fought most valointly
> At ye Water of Saugh,
> Along with Ledenhendry,
> Who did command the day:
> They vanquis the enemy
> And made them run away. (1707.)"

Adding the date was the finest touch of the whole. Ledenhendry, it seems, led the attack against the Cateran, and was getting the worst of it in a single combat with their leader the "Hawkit Stirk," when his friend Winter got behind the "Stirk" and hamstrung him. At all events, this is the country comment on the third line of the poem.

There were enough well-favoured Irish bullocks on Fat Haugh to raise two or three such Water of Saugh strifes; and in a field hard by were Warbler by Fourth Duke of Oxford, with Queen, Sultana, Lady Blanche, and another of Lord Raglan's daughters. There are not more than eight cows in the herd; and Canary, the dam of Confederate, the present bull, died just before her son won the reserve number in the yearling class at Stirling. The Cure was at Cortachy for a time, and it has also been a resting-place, temporary or final, for distinguished Shorthorns. Lord Raglan came here from Lord Kinnaird's when he was a two-year-old, and was passed on to Sittyton for £100. Young Ben, the winner of the aged bull class in '61 at Dublin, was fed off here, and so was the victorious Ivanhoe of the deep forequarter. The weary age of Rose of Autumn, who founded the Athelstaneford herd, found here a peaceful hermitage, when she had taken her second gold medium medal as a fifteen-year-old at Perth, and was sold to his lordship "at butchers' price if not incalf." Her days were nearly over, but the herdsman still dwelt lovingly on her "pleasant countenance" and her "grand braid hurdies."

Ere the family piper had done pacing the sward in front of the Castle, and sounding his *reveillé* through the crisp, bracing air, we had finished our stroll with the dairy. It is a pretty cottage building at the end of the flower-walk. The effect of the interior is all the more striking from the utter absence of anything elaborate. The milk-pans are of plain white delf, and the cream testers are marshalled, like the Commissioners on the woolsack in the Reform Bill picture, all in a row on the centre slab, and flanked by blue glass jugs and drinking cups. The curds and cream service stands on a rustic table in Lady Airlie's room, which looks into the dairy, and the plain cocoa-nut matting, the grey stone fender, and the cane-bottomed oak chairs with nothing to " give colour," except one scarlet leather back, all blend into one harmonious whole with the distaff and the spinning-wheel.

> "Kilve, thought I, is a favoured place,
> And so is Liswyn farm;"

but, always reserving Sutherland, give us an autumn morning's ramble through Penryn, Cortachy, and the Valley of the Hodder.

CHAPTER XI.
CORTACHY TO PERTH.

"Nothing is too good for exhibition-birds; give them daily exercise and an abundance of food. Linseed is calculated to give lustre to the plumage; and toast, soaked in ale, sprightliness, courage, and strength."
THE SCOTTISH HENWIFE.

Mr. Watson of Keillor—His Show-yard Career—Old Grannie—The Polled Herd Book—Experience of Southdowns—A Judge of Distance—Mr. Bowie's Herd—Dr. Murray of Carnoustie—The Story of The Cure—Major Douglas—Salmon Fishing on the Fife Coast—Speedie v. Seals—The Shark and her Whelps—Cock and Hen Salmon —The Carse of Gowrie—Rossie Priory—"The Scottish Henwife."

"THE castle of old Forfar" might once, as a poet observes, have been "*stuffit full of Inglishmen*"; but we had no time to inquire after the fate of our compatriots, as we pointed straight from Kirremuir to Maius of Kelly. Keillor, which has always been regarded as the very Warlaby of the "doddies," lies about twelve miles from it, and a little east of Cupar Angus. It will be four years come Martinmas since Mr. Watson left it, after a residence of four-and-fifty years, and retired to a new home in Perth. He was purely catholic in his cattle tastes. Bracelet, Charity, and one or two more of the pure Booths were the models he kept in his eye, in building up his blacks;

and even in a shire so strongly wedded to its own breed, he did not shrink from saying so. Many of his dearest friends lived over the Border— John Booth, Anthony Maynard, Wetherell, Torr, and Philip Skipworth—and he loved to go shorthorn and sheep judging with them to Ireland, and to call to mind Booth's merry jokes and his practicals on old Philip. He also had many "a quiet day at Wiscton" with the first earl among the shorthorns; and he was walking with his lordship on the race-course at Doncaster, just before Elis's St. Leger, when he first met Sir Charles Knightley. The old baronet began to rally him directly after they had been introduced, in allusion to the earl's politics, on "not keeping better company." Before the end of the week they met again at a sheep sale at Wooller, and for many years kept up a strong correspondence.

Old Jock (1), Strathmore (5), Angus (45), and Pat (29), were his four favourite bulls, and there is a strain of them in every great black herd. Old Jock was the most stylish of the lot, and showed, as his owner never scrupled to say, "much of the short-horn superiority in hair and touch." His son Pat thought nothing on one occasion of walking eighteen miles to a show, and winning; and his son, Hanton, made the herd fortunes of M'Combie, who bought him for 105gs. when he had won at Berwick. Old Jock was sold for 100 guineas, after taking a Highland Society first in 1844. In 1852 his son Grey Breasted Jock, or Second Jock, beat all the polled bulls in a sweepstakes at Perth, when he was thirteen; and Black Jock (3) and Young Jock (4) kept up the line.

"Keillor Watson," as he is always called, began to show in 1810, and won upwards of two hundred prizes for sheep and "doddies" in the next thirty-three years, principally at Strathmore (Cupar Angus), the Highland Society, and the Royal Irish. Some of these must be credited to thorough-breds and cart-horses, and among the medals and other trophies there are not a few race-cups. Old breeders still speak with rapture of the heifers which he showed at Perth in '29; and his Leicester rams were so good and level on that occasion, that each of the three judges had got a different one for first. 'Twenty-nine was also the year of his Smithfield heifer; and so delighted was Earl Spencer, the President of the Club, with her, that he requested that she should be modelled and struck off on a medal. He also gave the Irish a taste of his quality, and made several large sales there. His four-year-old Angus ox went over, and was placed first for the Purcell Challenge Cup at Belfast, and yet, strange to say, died after all in the plough at the Royal Home Farm when he was rising eighteen. Still his fame was in all lands, as a traveller in India found his portrait pasted up on a temple of Vishnu. His longevity was hereditary from his dam old Grannie, who gave no milk after she was 28, and ended in July, '59, a pilgrimage of 35½ years. From one to three she was often shown, and very seldom beaten as a cow; and her guardian, James Thompson, after forty-two years of service, received one hundred francs as a tribute from the *"Société Protectrice des Animaux."*

She is "the prima cow" of the polled herd-book, and dates from 1824; while Colonel, the premier bull, is six years her senior. This book, which was published in April, 1862, contains entries from 126 owners, 31 of them Galloway men. Of the 336 numbered bulls, 45 are Galloway, and the cows of the sort muster 95 out of 846.

Mr. Watson kept Leicesters on his low land, and southdowns, to which he had always a strong leaning, on the hill. In 1838, he could record that he had bred the latter for five-and-twenty years, that he thought them as hardy as the Cheviot, and that their snug-woolled heads and necks dried sooner after a storm. In another respect he found them very superior, as he could always fatten them much better off grass the year their lamb was taken from them. His experience of their hardihood was drawn from the fine middle range of the Seidlaw Hills, where they browsed upon the green sward, intermixed with whin and heather, five hundred to twelve hundred feet above the level of the sea, a spot "too high for Leicesters, and under the level at which the native black-face only thrives."

Scotch miles are proverbially long, although nothing to Shetland ones; and the day was far advanced before we reached Mains of Kelly. Mileage authorities differed so essentially, that after we had consulted an old woman, who overruled all previous decisions we put it confidentially to a lad with a cart, who was exactly wrong by one mile in the three. We thought the old lady was not within earshot during this interview; but she was, and we heard her indignant protest, "*greet lee*," in a voice quite beyond her years, and her word proved to be sound law.

Mains of Kelly lies about three miles from the flourishing little town of Arbroath, which has not fallen under the prophet's ban of being "a' dung down," like Dundee, and shows quite as few symptoms of it. Mr. Bowie's father came from Cockpen, three miles from Dalkeith, in 1809, and the son has held farms since 1834 under the lairds of Cockpen, or rather the Marquises of Dalhousie, The late Mr. Bowie had been an Angus breeder since 1809; and when his son began, it was on Mr. Fullarton Lindsay Carnegie's estate, with one of Mr. Colvill's Boysick tribe. Cupbearer by Pat was the junior's "opening star," and the white inside the hocks and in the centre of the tail has always been the Bowie coat-of-arms. As he also bred Hanton by Pat, a bull of great touch and size, he could virtually claim the first and fifth honours in the great Paris contest. How matters were conducted by the jury, both there and at Hamburgh, will always be a problem to the breeders, who got quite confused with what they still term "the showing of hands and measuring of tails."

His best stock on the male side are entirely due to Watson of Keillor and Fullarton of Ardovie. The dam of Angus came from Keillor, and did him rare service. When she had not a whole tooth in her head, and had missed a year, she presented him with a calf in January; and then she had milk fever, and, after all, brought up four more, two of which, Second Earl Spencer and Cupbearer, were first and second-prize bulls at Berwick. Second Earl Spencer was her own calf, and when he was killed, at four years old, he weighed nearly 120 Dutch stone of 17½lbs.

The herd numbers 80 to 90, and none are fatted off except those bred on the spot, and generally at 2½ to 3 years old. Yellow swedes and oatstraw are their principal fare, and " good and grey" potatoes at the beginning of the season, when the price of the crop will warrant it. They are generally bred more for use than sale, and Mr. Bowie firmly believes that he "can keep four for three shorthorns." Occasionally he has tried Galloway crosses, but it never seemed to hit nicely, and the calves came coarse. He very seldom brings out any cows or heifers, except at the local shows. Of late, his great bulls have been his Kelso winners, Jim Crow and Tom. "Jim," the two-year-old winner of that day against Julius Cæsar and Commodore Trunnion, had a rare flank and quarter, but not such a back as Tom, another grandson of Hanton's, who made a very fair fight with Fox Maule in those Springwood Park meadows. Tom is a very lusty bull, and, as it was said of him at

Kelso, "a whole ox before the shoulder." Old Lady Anne, with brown hair and white marks, was nursing her calf in the byre. The last Kelly relic of Queen Mother, who came here for £40 when she was nearly worn out, and had one calf after, is Victoria by Cupbearer, whose progeny promises well. Lola Montes, the dam of Charlotte, also migrated from Tillyfour in her old age; and she, too, left one pledge behind her, the last sheaf of a very rich harvest. Mr. Bowie had a sale in 1857, when two dozen store animals averaged £37 0s. 7¼d., and twelve bullocks from two-and-a-half to three-and-a-quarter years £30 16s. 8d. Caroline, the dam of Clarissa, went at 67 guineas to the Earl of Southesk, Standard Bearer, a twenty-two-month bull at 89 guineas to Mr. M'Combie, who bought Hanton from this herd, and the two dozen were dispersed into all counties from Essex to Inverness.

To go to Carnoustie and not call on Dr. Murray was a thing not to be thought of. We had an interview with him in that remarkable room, on whose walls hang Voltigeur, Sir Tatton Sykes, and Refraction, and a photo of The Cure when he was down in the world. David Bonella, late groom and surgery boy, holds the son of Physician in stable guise; and the Doctor, who has a fancy for the camera, has also introduced him at the head of one of the earliest Cures, a filly from Lily Adey. A pile of *Ruffs* was on the drawers, and he consulted them at intervals, sitting in his shirt sleeves at the edge of his bed. Save by one other man, who asked us to recommend him a prophet, we never heard the name of a race-horse breathed before we got to Perth.

If his countrymen were dead on the point, Dr. Murray completely atoned for them. The history of The Cure was the first great subject, and he spake of the bay on this wise: "Mr. Rait got The Cure from Lord Airlie. He had been a year in England, and got Lambton from Elphine by Birdcatcher. While Mr. Rait had him he hadn't a thorough-bred mare, and only three or four half-bred ones. They tried to make a hunter of him, but he wouldn't jump over a stick, or these slippers of mine. He gave them some desperate falls, and broke a groom's leg, so he was sent to Mr. Jones, and then he came to me. He was offered to a veterinary surgeon gratis, but he declined him; so I said to Mr. Jones, '*I'll give you three half-crowns for him.*' I had never seen him, mind you; but I knew all about his running. So he was sent over. I saw a pony thing coming up the brae, and I said to myself, '*That round Highland pony of a horse win the Champagne, and run second to Foig-a-Ballagh for the Leger.*' There was quite a revolution in my views of a race-horse. He was lame on the off foreleg. Thinks I to myself, 'He's only ten or eleven; I can surely mend him at that age;' so I paid my 7s. 6d., and I blistered his leg. Mind you, the three legs which wern't lame had running thrushes, which I healed. The leg became fine after blistering. I never use bandages.

"Well, faith! I began him on the road. I found him a little lazy, and I never could bear a lazy horse. Then I bought a thorough-bred mare at Lord Panmure's sale—Lily Adey. Time went on, and his son Lambton's exploits began to make a noise in the world. Lambton won eleven out of twelve races, and people began to say, '*Where is The Cure?*' So I sold him with three legs and a wooden one for £50, and a filly and £25 more if he was sold for £200. I was glad to get rid of him. There were two law-suits behind him. He was once claimed from me on a race-course by a trainer, who said he had given £10 for him. He was fifteen hands when his feet were pared, and, to my astonishment, he was fifteen-three in the next advertisement.*

"I have measurements of him on every point, and I'm curious on points. He was 5 feet 10¾ inches in his girth when fat, and 5 feet 7⅞ inches when he was in moderate work, and 5 feet 5 inches during my process of training. I trained him myself, whenever my practice allowed. I always trained by time. Lily Adey or her foal led the work, with my surgery lad up. The Cure took lots of work to reduce him; he was as round as a bullock. He had perhaps two or three hours' training between seven and three—short gallops, and not clothed. You get more out of them without clothing. He never broke down with my method. Faith! The Cure never tired, and never had a blanket on him.

* Mr. Osborne assures us that The Cure is 15-2 in his shoes.

I had a faster one than either of them—North Star, who won two or three heats at Newcastle and Lancaster. I measured his stride on the sands, where it is always rather shorter than on turf. It was 17 feet, and 18 feet by urging, and The Cure's only 16¼. Nuthook's was 17, and The General colt's 17 feet 7 inches.

"They did their regular work on Monifeith Links, between here and Broughty Ferry. It was once a racecourse, but now it's cut up with rabbits. There are Barrie Links near it, and some stables of the late Lord Panmure's on it under the wood. His horses were there once. John Howe from Newmarket trained there. I thought six weeks or two months quite enough to train The Cure or any other horse. He soon took the ditches nicely enough in my practice: he took the little ditches fine. I raced him, did I ? I put in a first appearance with him at Perth. Davie, the lad, rode him, and he was third and last to Haricot. The saddle turned round, and Davie fell. I heard the cry, ' *The boy's off!*' I leapt the Stand rails, and a trainer's horse knocked me down. When I came round, I found myself in the weighing-house. I had got my cheek-bone broken, and a black eye. ' *Did I take The Cure, that night when I got home from Perth, to the crying wife?*'—' *Na! it would be Lily Adey, I'se sure o't.*' After Paisley, I rode him myself at Stirling, and took him off after the second heat. I thought they would seize him. I used to go to the races with my toggery under my coat and trousers. You can see it all for yourself very pretty in *Ruff's Guide*, while I get a little mint for a lassie there."

And we saw it, sure enough: " Paisley, October 9th, 1851; Paisley, August 6th, 1852; and Stirling, September 2nd, 1852," when the horse was ten and eleven years old. We read the name of "*Mr. Murray*, 9*st.* 11*lbs.*" attached to the last race; and the Doctor seemed quite astonished. "*I never got to that weight in my life;* and I never declared over weight. I was nearer eleven stone; but they're not very particular at Stirling. Then I bought Lambton. The Cure filly from Lily Adey was second to Heir of Linne for the Queen's Plate at Musselburgh, and beat Yorkshire Grey. I ran Lambton two or three times, but he was no good; so I hacked him and I have a filly by him. Then I had North Star by Arundel, a rare one for speed, and the quickest at it, but a bolter. It was at Paisley he bolted among the crowd at the Grand Stand. I felt my feet rapping on their heads for all the world just like when you run your stick along an iron railing. He jumped at a ditch ten or twelve feet deep, and got his fore feet on the other bank, and played plunk with me to the bottom among moss and peat. I heard them shout, '*He's drowned!*' I got out fast enough, and they got ropes and pulled out North Star. Faith! he was a black horse when he came up. I never went back to weigh, but I just bolted from Paisley by the next train. There was a paragraph in the papers

R

next day—'Accident at the races: a man's leg broken, an arm dislocated; taken promptly to the infirmary'—all that sort of thing, and just through this abominable bolting. He fell with me again, and put out my thumbs, and broke two of my teeth. I finished him at Perth. I knew he would bolt at the gate if he could; so I kept the horses between me and it, and when I went up to cut them down, he was round in a circle, and we were down in a heap. I up, and I'd have caught them, but a lad said: *'Look! his leg's lolloping about.'* I had him shot there and then; his leg was broken under the knee into twenty pieces.

"Lambton was no good, and I gave him to Jemmy Laing, and he went back to England. He got his legs under the van partition coming back from the races one day, and I thought he was a dead horse. His wrestling to get his legs out produced exostosis on both his hind pasterns: his forelegs were bad enough before, and now he was a cripple all round."

Such are the strange Scottish antecedents, not only of the little bay horse who was second to Voltigeur in a field of sixteen sires for the 100-guinea prize at Middlesboro,' the sire of Underhand, and the now dearly beloved of John Osborne; but of one which has got, perhaps, more winners in proportion to his chances than any horse of the day. The Cure was a year at the Royal Stud, and left a 1,000-guinea yearling behind him, and Lambton headed the Doncaster yearling poll in 1864 with "the Prince of

Wales' 1,080 gs." There must be some virtue in medical superintendence or Carnoustie air!

On our way to Dundee we passed the Links, on which The Cure and Lambton took up the tale, when King David, Bustler, Ledstone, Harlequin, and the other sheeted tenants of the Panmure Stables had run their course. The old "Cock of the Glen," Major Douglas, once the Osbaldeston of Scotland both with the trigger and in the saddle, gave us a kindly greeting as we passed through Broughty Ferry. Black tan is the Gordon setter colour,* to which the Major still steadily holds, and a beautiful troop of them were at his heels, with one or two fox terriers, descended from Rage of the Rufford. With a nervous horse, the ride from the Ferry to Dundee is a peculiarly difficult one, as the railroad cheeks the road all the way, and consequently our time seemed to be occupied with perpetual bursts over clover in the open. However, we were among the forest of chimneys at last, and Mr. Speedie of Perth was ready for us on the pier.

Anything for a change after so much beef and

---

* *The Field* gives the following account of the alloy in the Gordon Castle setters: "Some time about the year 1826 there was a celebrated sheep-dog belonging to a shepherd who lived far up on the Findhorn. Among her other accomplishments, the shepherd, being a bit of a poacher, had taught her to find grouse, for which she had a wonderful gift; she knew by a wave of the hand and a word whether grouse or sheep were wanted. When she had found grouse the shepherd would say a word or two to her in Gaelic, go down the hill for his gun, and on his return find the bitch still watching the grouse : it was more like watching than regular pointing ; you might have fancied there were sheep in front of her to be looked after. The Duke of Gordon (then Marquis of Huntly) heard of this bitch, and begged her of the shepherd. The shepherd unwillingly gave her to the 'Cock of the North.' The marquis put her to one of his best setters, and some of her first litter were black-and-tan. She herself was long, low, rather smooth for a colley, and black with very light tan."

mutton; so we had settled to have a day with him among his salmon nets on the coast of Fife. Some caravans bound to the Dundee Fair had to be wheeled out of the packet-boat, and she was soon returning to Newport, and its river bank of "Pluck ye Crow." Mr. Speedie can stand by his nets like a Stoic, and see shot after shot returning void on a long summer day; but that seals should venture to claim a dividend of the salmon pricks him to the very quick. Some of them have had the temerity to venture up as far as Stanley; but "Cowie Jock" was on the look out for them after that feat, and impaled no less than four at Higham Bank, half-way up the Tay, with his spikes, in one tide. None were visible off Dogger Bank that morning, or farther away among those dreary wreck ribs; but, "*There's Cowie!*" said his natural enemy, as we drove across Drumley sands. There he was sure enough, with his black head bobbing up and down in the ebb tide of the channel, and chaffing Captain Maitland Douglas's fishermen.

One of them, 50 stone in weight, was taken out of the salmon trade not many months before, and in his rage at losing his licence without any compensation, he bit the foot-spar of the boat in twain, as if it were a tobacco-pipe. "*Fifty pints of oil came out of him, the rascal, and the men got £3 for it.*" They chase the salmon into a chamber of the net, and then swoop down on to him like an eagle, and fish him out with their claws. An hereditary craftsman at the trade tosses his fish in the air, catches him again, and, after three bites, throws him away contemptuously, with just the bare back-bone attached to the head. Four large salmon, weighing very little short of 100lbs. avoirdupois, were taken by one in an hour before Mr. Speedie's very eyes, so that he may well place a guinea on the head of a " 70-*stone black rascal*," and vow "*to make him out next year if I'm spared.*" Rifles are of no avail unless you hit the brain; and one great fellow was taken at last, full of conical and spherical balls, which had merely come to grief in his fat. A seal net, with a salmon dangling from the top of the chamber, is the only sure way of catching them, and it takes eight or ten men to work it properly.

The ooze was one orchestra of sea-birds. "Larks of the woodcock tribe," with brown backs and white breasts, were skimming along over the shallows, without any definite object, like gentlemen "unattached." Sea pyats were mingling their shrieks with the low cry of the curlew; and the sea scarts were the busiest fishermen of the whole, and screaming out their protests in chorus, when the hawks, "who like the other lads to fish for them," pounced on them with the dash of a Semmes, and made them hand over on the spot. Thus the Speedies and the sea scarts are equally tried in this life.

This great fish dealer has 200 men in work, and upwards of twenty stations on the Fife and Montrose coasts, as well as on the Earn and Tay, and his rent for the present season is considerably above £9,000. The Fife chain of stations begins at the bend beyond Drumley, and goes down nearly to St. Andrews. The stations are on the Muir, three or four hundred yards from the shore, and at each of them he keeps an overseer and four men. The old building with the low door, the earth and heather on the top, and a load of boxes with cord handles in a pile beside it, looks like a cave of stalactites when you first enter. Then as the light breaks in, a great cock-salmon's beak or grilse's head is found protruding from the rough ice; and when the overseer looks at his book, you may hear that "96 salmon and grilse were taken yesterday." They had just taken something more than they expected in a shark, which had three brace of young ones swimming after her, and she whelped ten brace more on the grass before she died. She would have made the fortune of a caravan, but there she lay neglected among the sedge, with her green back and pale-slate belly, and all the little things around her; and her captors only remarked that she had a three-year-old mouth and a sand-paper hide.

Trying the nets was the sight of the day. The man climbs along the side rope, and takes the fish out of chamber after chamber, and slips his cord through the gills, and at last he descends from his perch, and wades out, dragging after him a regular bouquet of all weights from 40 to 5 lbs. Sometimes they are all cocks or all hens, and on the rivers, especially during September and October, hens get very voracious when they are near spawning, and take the fly much more easily. The difference is so marked that, towards the latter end of one September, Mr. Speedie kippered ninety hens, and only two cocks. The fishing is very good at Tent's Muir when the east wind blows right into the shore; but there has been no vintage like that of '30. In the sea they always go before the wind, and in rivers they swim right into the wind's eye, like ducks up a decoy.

But we have looked over all the fishing stations, and explored the old nets in the granary, which are to be sold into the paper trade, and we are once more driving over the long stretch of hard sand towards Eden Mouth, rich in spotted trout and mussel beds. The black slug and the mussel scarp are quite a subject of dispute between the solan geese from the Bell Rock and the St. Andrews fishermen who take them off to Aberdeen and Peterhead as bait for haddock and cod, and all the other treasures of the deep-sea fishing.

There our fishing ends for the day. We have no time for St. Andrews, that fine old city in decay, with its Cathedral and Palace of Cardinal Beatson, and the colleges to which thousands of students— Campbell and Chalmers, Ivory and Leslie, Leyden and Milne, Playfair and Ferguson—journeyed so reverently in their time. That gorse on the hill, so dear to the Fife, is passed in our homeward ride; so is the jaunty beaver with the twig in his mouth, over the hatter's shop in Dundee; and we are at last in the Carse of Gowrie, and exceedingly thankful that we are not a wood pigeon.

The Carse proper begins at Invergowrie—the scene of the first preaching of John Knox—some five miles from Dundee, and extends to Kinfauns, where the Tay narrows at about the same distance below Perth. "The braes" run down from the Seidlaw Hills, and the whole of that great wheat plain below was once supposed to be under water, and dotted with divers "Inches" or islands. Leases for life and wheat were still stronger traditions, but the former have nearly all merged into nineteen-year ones, and beans, barley, and oats have found their way into its Farmer's Calendar. In 1827 farmers laughed at the notion of permanent grass and turnips. Lord Kinnaird first laid down the former on the braes of Rossie, and Charles Playfair of Inchmichael took his stand on turnips, and fed a bullock as well. Nearly all the turnips are pulled and taken to the straw yards or byres, but the system is gradually creeping in of eating them off the ground with sheep. Still stock is not the *forte* of the Carse. It was the high prices of '96 which first brought it out, and the glistening furrow and the white crop have been its glory ever since.

Owing to the absence of hedges, there is rather a bare primitive look about it, but this disappears as you begin to ascend the braes of Rossie, among the

beech and thorn hedges which lead to the Priory.
The wooded hill behind was planted by Lord Kinnaird's grandfather, principally with fir cones from the Forest of the Mar, interspersed with beech, oak, and elm, which were just beginning to wear all the varied hues of September. His lordship farms about 1,200 acres, and has also three farms temporarily in hand for drainage and deep ploughing by his Fowler and Howard ploughs. His flock of breeding ewes is kept up to 300 Leicesters and 150 Oxford Downs or rather the dun or grey-faced Cotswolds. The Leicesters date from '36, and come from the orthodox Midland combination of Stone, Burgess, and Buckley, with Sanday and the Borderers to follow. At one time he bred Southdowns, but Dundee liked something fatter. The Leicester wedders are fed off at from fifteen to eighteen months, and the tups are sold for breeding. Some of them are reserved to cross the Oxford Downs, of which his lordship buys drafts, as well as rams every year from Clark of William Strip, and Smith of Bibury. Most of the lambs are sold from the teat, but 240 Oxford Down wedders are put up under cover each year. The lots are fed for about twelve weeks each, principally on turnips, bruised oats, and cake, and the mutton is sent to Edinburgh and Glasgow. The turnips are washed below, and sent up a hoist on to the level of the sheep pens, which are all on sparred floors, to be pulped. Nineteen or twenty firsts have already rewarded the plan, and the card for the best pen of any age or

breed at a leading Irish show is among the white, blue, and orange array. Sixty to seventy bullocks are also fed off annually at Mill Hill, where his lordship's factor, Mr. M'Claren, resides. The feeding is principally managed in boxes, under one roof, and the partitions take out so that the dung-cart can be backed in. His lordship considers, from his experiments, that the value of manure thus made, and ploughed in at once, is very superior to that made in the usual way. The pigs are from Hewer and Tombs of Hatherop. His lordship began with Berkshires, and has come back to them as most prolific, and most easily fed, after running over all the white-pig gamut, from the large Yorkshires and Radnors, so on through the three W's—Windsor, Wenlock, and Wiley.

Mill Hill has also a cottage fitted up as a Turkish bath for cattle, which has been found most efficacious in cases of pleura and rheumatism. The time for keeping them in is 1 to 1½ hours, and they are well douched with cold water when they come into the drying parlour.

The Castle Hill steading, where young William Ward was the herdsman, lies about a mile and-a-half from Mill Hill, on the Perth side of the Priory. It is now twenty-nine years since his lordship entered upon the breeding of pure Shorthorns. He began from the stock of Rose of Cottam, which he crossed with Lord Ducie's Champion (11264) and with Mr. Fawkes's Belted Will (9952). He then hired

Prince Arthur (13497) and Prince Oscar (16757) from Mr. Richard Booth; and Rossie was also one of Lord Raglan's resting-places, as he gradually moved up North from Southwicke and Athelstaneford, and so on, after three years in the Carse, to Cortachy and Sittyton. Prince Alfred could not be got at the time, and Lord Privy Seal (16444) was bought from Hill Head. Cherry Duke 2nd is the only Bates bull on the list, but when his gold medium medal had been claimed at Stirling he was very smartly sent to the fleshers. Lord Privy Seal is one of those compact, nice, little bulls, who looks as if, to adopt an expression of Mr. Wetherell's, *" he had been put into a lemon-squeezer, and just made the right size"*; and his cross with a big red cow, Jenny Groat, now at Mr. Creighton's, near Inchture, produced the well known Great Seal (19905). Lord John Russell (16417) brought the gold medal for the best bull in the yard back from Belfast to Rossie, and it was Lord Privy Seal's lot always to be second to him. Prince Louis, a winner of six first prizes, had just been sold to Mr. Milne; but Lord Louis, a Kelso bronze medallist and third at Stirling, and Grand Royal, who was bought from Mr. Torr, were in the well-filled bull ranks, and so was Baron Highthorn.*

Ventilation is a great point with his lordship, and it is well managed by pantiles, so as to secure good

---

* Since our visit, Mr. Carr's Prince of Windsor has been purchased, and Lord Privy Seal exchanged for Messrs. Cruickshank's Windsor Augustus.

temperature without a draught, at the top and through the ledges. In the calf-box range we found a young white Breastplate, and a bit of Bates in the shape of a roan by Lord Oxford (20214) from Mr. Grant Duff's Louise tribe, of which Louise 2nd by Champion (11264), "the old original bull of all," and Louise 3rd by Lord Raglan seemed the leading pair. There were several boxes ten feet square for cows and calves; and Stumpy, the prize Ayrshire, was consorting in a row of stalls with Sister Mary by Sir Colin (16953), and her daughter Sister Ethel by King Egbert (18134) who was "still more Booth in her head." There, too, were Lady Gertrude by M'Turk (14872), Princess Laura by Prince Arthur, and Maid of Orleans by Lord of the Valley (14837) and true to her sire's horn; and we can well see the truth of what Ward says: *"It will soon be Branches over again, for his lordship's all for Booth."*

And so we leave the shady groves and sunny slopes of the Priory, and ride on four miles past " Patience on a tram-road," and carts bearing tons of Dundee police manure, to Inchmartine, the home of the celebrated Henwife of the Carse. There is nothing like approaching a subject gradually; and certainly we did so, as *" a little waterhen crossed the drive into the laurels"* was our first note. Mrs. Fergusson Blair's love of hens is scarcely twelve years old, and it had its origin on board a steamer from London to Edinburgh. Looking at some coops of Cochins helped to beguile the weariness of the saloon, and two of the hens were purchased for ten guineas. They were not worth the money, and none of the blood is left. Stretch of Liverpool and Chace of Manchester supplied new strains, the latter buff and white; and Mrs. Blair also examined the different Zoological Gardens at Paris and Antwerp, and bought a few pens at £4 a-piece. The Rev. G. Hustler furnished her first Dorkings, but they have been replaced by Captain Hornby's and Lewry's breed. The Bramahs, light and dark, white and cinnamon, came from Tebay and Miss Watts. The old Scotch greys had only a short reign, and were presented to the Emperor of the French in the spring of '62, six years after Mrs. Blair had led the way in importing the Creve-Cœurs, with their comb like a split heart, from Normandy, as well as Houdan and La Fleche. Polish never gained much hold here, but bantams flourish under the respective heads of black, white, and game. There were a few Spanish, but, hidalgoes though they be, they have all been banished to the farm. Some of the Turkey patriarchs came direct from America; and Fowler's Prebendal farm which furnished the first Rouen ducks and geese to Inchmartine, now adopts the principle of amicable exchange. For a dozen Dorking eggs the charge is from two to four guineas, and for others in proportion; and when Edinburgh could not furnish Holyrood with a regular supply during Prince Alfred's winter stay, the purveyor fell back in his need on Inchmartine.

Mrs. Blair manages the whole of her son Mr. Douglas Allen's estate, and keeps 160 acres in her own hand round Inchmartine. Wet or dry, summer or winter, she never omits her two o'clock round to her poultry yards with two baskets of rissoles. These are called oatmeal by courtesy; but a number of ingredients—buckwheat, linseed, spice, and pimento —are all veiled under that title. Chamberlain and Penn's food is used largely, and so is aromatic condiment; and old ale, bread, potato, chicken, and other dainties have got into that wonderful bowl which is devoted to the clearings of the dining-room. Wheat, barley and Indian corn are the staple of the out and in-door relief, which the girl and the man who act as sub-overseers under "Annie" the head woman, deal out twice a-day. They have full employment, as there are sometimes fifteen hundred head of fowls of all kinds and in all stages. The setting hens have to be duly lifted off their eggs, and put out for half-an-hour to exercise; and "the sad vicissitude of things," from a cock catching cold to the chances of egg-roasting in the eccalobion, demands the most careful surveillance.

Broom, a first-prize silver grey, is as good as a sentinel at the front door, and his taps at the window a few minutes before two o'clock became loud and frequent. He is entitled to the most unbounded licence and consideration, as Sydenham, Liverpool, Edinburgh, Glasgow, Perth, and Paris have all showered their highest honours in turn on that now hoary head. He is five years old, and therefore long past his Dorking prime. Silver greys have no chance against the heavy greys, as at his best he never turned the beam at 9½lbs. His plumage is perfect —the jet-black breast and flowing tail, the dark and distinct blue bar in his wing, the silver hackle, and the straight indented clear-cut comb. "Greedy" and "Missy," are the companions of his fortunes; and their beautiful ash-coloured hue, with the white feather shafts and the robin red-breast, have often led them to victory. "Broom" conducts us most courteously to the edge of his domain, and there leaves us among the pencilled Bramahs. Mrs. Blair has composed a new strain in Bramahs of a composite colour between the lemon and grey, very suitable for the farmyard and for maternal duties, which they perform à merveille.

The stable has two beautiful thorough-bred blacks by British Yeoman, bred by Miss Bell of Woodhouselees, an old fellow-labourer in the poultry cause, and the glass-house in the stable yards seems a nursery for valetudinarian and motherless chickens, or rather the late and the delicate. The " Laurel House" is quite a step beyond, and is, in fact, a fashionable seminary for pullets of all breeds before they are introduced. The master of the "Wood House" is a heavy grey Dorking, a first both at the Sydenham and Birmingham shows, with four hens at his side. "Growley" has also made himself a name at home and at Sydenham; but he takes his ease now in a hollow under a tree hard by the garden walk, and is insatiable in his basket cravings when his mistress appears. A hen and her chickens sit under every tree in the orchard, but they are not brought there till they are past coop estate, and the coops have been lime-washed and put away for another summer. "The Monastery" at the bottom of the orchard is a sort of mysterious penthouse, with rows of perches on each side, and tenanted at times by upwards of four hundred cockerils, from which the future public characters and private sale birds are selected. The rest are either eaten or sold round home, or put into the Edinburgh winter draft. "The Nunnery" is also composed of rafters, and contains pullets of every tribe and kindred from one to five months.

Here Annie in her white crazy appeared upon the scene, and confessed her love for a little cannibal of a Crève-Cœur, as black as a sloe, which had just attempted the life of an unoffending Chamois Pole. She loved it for its very mischief, and when she had called it *"just an impudent little smout,"* she kissed it fondly to make amends, and added in dark speech that it was *"feeding just like a little linty,"* and *" was able to keep its ain part."* Even long after roosting time her thoughts are with them, as she and Kitty the girl knit, and roast the oyster shells. The green, save and except the croquet-ground, is entirely given up to chicken-coops during the season; and some of the coops are glazed Crystal Palace fashion. The other houses are of quite an unpretending order;

and behind them the boar resides which beat Wainman and Dickin for the first prize at Stirling, as the orange card on his van testifies. Little Partridge-Cochins receive Annie's assurance of being "*Wee Petties;*" and she tells the story of a sick cock which has been washed with hot vinegar and water for an attack of cold, and quite "*enjoyed its castor-oil and a pill.*" Some little chickens just hatched are sharing the warmth of the poultry cowhouse with an Ayrshire, the excellency of whose strength principally goes to the fowls in the shape of curd. There, too, are all the show-baskets lined with pink calico, and ready for the next campaign; and the dainty commissariat department is full of Indian-corn, oats, oatmeal, sharps, and sacks of wheat *galore*, with lettuce-cabbage and onion all ready for the chopper. On the ladder leading to the loft sat "Mussie," an aged bantam. It is getting quite "doited" in its head with infirmity, but Annie perceives it not. With her it is still "*a little wee monkey as happy as you like*," and it gets bits of egg as a solace from her breakfast, and beef from her dinner.

But Annie has not much time just now to spare for endearments, as she is in the society of "Smith," the cock turkey, which never condescends to feed off the ground, but only out of hand, dish, or basket. Mrs. Blair "hoped" he was 38lbs., and he must certainly have been pretty near it, thanks to a loaf of bread per diem. There are no whites among the Inchmartine turkeys, which are principally of American blood; and their nurseries are situated in some of the old grass avenues, apart from the chickens. Turkeys are rare mothers, as they never tread upon their young ones; and to them and the hens the greater part of the goose eggs are confided at hatching time. Each goose lays about fifty eggs if they are taken away, and thirty if they are left; and hens generally take three. Their maternity cares, under such circumstances, are rather chequered, as the goslings give the hen the cut direct during the day, and then creep in beside her at night. They are principally of the grey-imported Toulouse breed, and some of the best have come from Viscount Clari, the Emperor of the French's cousin. One gander was up to about 28lbs., and the Birmingham pen of goslings was "framing for 66lbs."

They are as good watch-dogs as their Roman ancestors; and the moment Mrs. Blair was seen, they came like a mighty, rushing wind from the other end of the paddock, and it was a mercy she could keep any shawl on her back, when she became the object of their concentrated tugs. The Rouen ducks were there some forty strong, with the green head, white neck-ring, rich claret breast, and blue ribbon on the wing; but the draft mark was relentlessly set against one of the finest drakes for the very dark-green shade of his yellow bill. The Dorkings lay in the woods, but come home to roost in houses with very low perches, which are specially provided to meet their tendency to becoming bumble-footed with years. Old Jack is spending a gouty but a happy old age, and has retired with three hens nearly as old as himself, who have had unlimited ale and beef in their time. Annie of course embraces this very Abraham of Dorkings, and assigns "*hundreds of chickens off him*" as her reason for the act. The gold medal at Paris was one of his trophies, and he has paid so many visits to the Crystal Palace and other shows, that at last he learnt to love this vagrant sort of life, and still comes solemnly to the side of the show-baskets when Annie and her staff are packing up his juniors. The silver-grey Dorking, "England" is the paradox of the place, as he never crossed the Border in his life. The topknots of the Buff Poles fairly drown them; and among a lot of good white Cochins, there is one with a vulture hock, which is condemned, as Mrs. Blair is very rigid in her observance of poultry rules, albeit some judges will pass it.

The pencilled and white Bramahs, a combination of her own breed with Priest and Tebay, seem more after her heart, with their black necklaces, broad, black or spotted breasts, and their fine, dignified carriage, on those short and well-feathered legs. They are excellent mothers, and it was quite the treat of the day to see their salmon-pink eggs handed into Mrs. Blair's basket, which reminded us of Pache Egg Monday, with its store of every hue. The Birmingham first-prize Crève-Cœurs were strutting gallantly about, as if they heard the words of their mistress that they are "the best table fowl in the world." The egg chambers of the La Fleche know little rest, except at moulting time, and therefore with a special view to egg profit Mrs. Blair's farm is principally stocked with them. *"Dinna fret yourself, Cocky,"* says Annie to one of them as she gathers it up to look at its feet; but it is on them again in an instant, as she snatches up a wand, and rushes best pace to the other side of the green. A three-cup Bramah has demeaned itself so far as to quarrel with a nameless Dorking; and as we retire through the evening shades, the last words that float on the breeze are, *" Oh! ye're a bad boy, fighting !"*

CHAPTER XII.
PERTH TO DUNKELD.

"A teering nag was Sharpe's Canteen,
　But nowt to old Springkell;
And Chassé was a ganner,
　And Inheritor as well:
A tougher meer than Modesty
　The Border never crossed;
But Cumberland just banged them a'
　Wi' Ramshay's Lanercost."
　　　　　　　　CUMBERLAND BALLAD.

The late Lord Lynedoch—Net Fishing at the Lynn of Campsie—Mr. Speedie's Ice House—Salmon Prices—The Scone Steading—Mr. Paton's Gun Shop—The Caledonian Hunt Club at Perth—The Club Rules—Its different Places of Meeting—Forty-five Years of Racing.

DIVERS rambles down the wooded banks of the Almond, with our old friend "Hawthorne," beguiled the next two days. Of course we visited the graves of Bessie Bell and Mary Gray, who live and move again as Clydesdale pairs in many a Scottish show-yard, and the ruined cottage of the late Lord Lynedoch. The veteran died at 93, but to the last he hated to be thought old, and would go aside with his servant, that no one might see him helped on horseback. When he shot grouse at a very advanced age on Glenesk, he could not be persuaded to sleep in the lodge, but

"Lay, like a warrior taking his rest,"

on his own iron bedstead and bear rug, in a portable

tent outside, and caught such a violent cold that he had to return home. He loved hunting, and went a great deal with the Duke of Grafton's during his residence at Cosgrove Priory, in their Stony Stratford country. The cottage at Lynedoch, where he spent three months of the year, has never been restored since his death. It stands on a knoll commanding one of the finest views of the river, and is a mass of beautiful decay in summer time, with the birds carolling on its yellow-green thatch, and the creepers clinging to its ruined trellises. A black retriever paced solemnly up and down the old passages, and in and out of the sashless drawing-room windows, as if it were the *genius loci*, and the flower-garden and vinery formed a rich prairie for rabbits and young pheasants.

There was not much to be seen on the Inches of Perth. "Good things" may have been "landed" there by racing men, but scarcely a salmon came to the net that morning. The Lynn of Campsie, five or six miles higher up the Tay near Stanley, had very different sport to show. It is close by Catholes, which is a very famous place for the rod. We had to be ferried across from islet to islet, among strange granite ridges and mountain ashes, and furze, before we reached Mr. Speedie's head-quarters at the Lynn. The men were taking a quiet smoke after a shot, with ten or twelve salmon at their side, when one came with a dart and a curl out of the water, 200 yards below, at a fly, and in an instant they were in their boats, and nailed him or a kinsman. *"Slacken the net,"* said Mr. Speedie, as the sullen splash told that a big fish was in its toils; but the order was just a moment too late, as he had gone through the meshes like the 20-pounder which he was, and up stream, without any more thoughts of flies. The nets were soon righted, and the misfortune as well, as the very next shot produced another fish and rather the bigger of the twain. Three shots in an hour where the stream runs rapidly, as it does near the falls, is very fair work, and the number of fathoms of net which they work is less or more in proportion to the rapidity or sluggishness of the stream. A southerly-west wind, a cloudy sky, and dark water are the three great essentials of a favourable shot. The men work night and day, in detachments of six or seven, twelve hours each, and on the Tay their season lasts 178 days, from February 1st to August 26th. Why their open time is longer than the other rivers Parliament alone can tell. From 6 p.m. on Saturday till 6 a.m. on Monday the river has rest from nets, as well as for the other 187 days in the year, during which the men watch and weave nets. Their fishing pay is 3s. a day, inclusive of boot money, and they get the old nets and fish occasionally as their perquisites. Eighteen pounds is about the salmon average; grilses range from 6 lbs. to 12 lbs.; and the smoults increase in their progress towards salmon estate at from 6 lbs. to 10 lbs. a year.

There were nearly seven hundred salmon and grilse, the produce of all his stations, in Mr. Speedie's fish-house, but he has had as many as a thousand odd. They are as bright as silver when they are fresh run from the sea after a spate, but as the season advances they get more chameleon-like, and take the colour of the stones or gravel where they lie. Before the railroad days, they used to go by the steamers from Dundee; but now the fish train leaves Perth each day at two o'clock, and reaches London at four in the morning. The consignors of course pay the carriage to London, and are charged 5 per cent. commission on sales. Mr. Groves is one of the largest London buyers of Tay fish, and Edinburgh and Glasgow do a great business as well. Middle men also purchase them in London, to repack and send on to France. They leave Perth in 200lb. boxes, with 60 or 70lb. weight of ice upon the top of them; and Mr. Speedie, who has sent off forty-four boxes a day at times, uses no less than 700 tons of rough ice in the season. Without ice they will keep pretty well for two days, and then they begin to go rapidly at the gall; but covered up in an ice-house they will be as good as ever at the end of a fortnight. In spring the heavy salmon of 20lbs. and upwards sell best, and fetch 6d. a pound more, as the west-end parties are larger, and better cuts are required; and by the end of June the middlesized get the run. Prices are seldom the same for two successive seasons. In 1863 and '54 (the cholera year) they began at 3s. a pound, and went down, from perfectly opposite causes, as low as 2½d., whereas in 1864 they were never below 9d.

A stroll of a couple of miles over the bridge, and up the opposite side of the river, brought us to the beautiful new steading of Scone. The Star, the Crescent, and the Thistle are on its gable ends, and above one of them stands a small image of St. Andrew, with his cross on his breast, and his "haughty motto" encircling him. The system of half-open yards is not pushed to nearly the same extent that it is at Lord Southesk's; and the details of the whole steading, differ very widely from its Perthshire rival at Keir. The open yards are in the centre, and communicate by a portcullis gate; and the pigsties, stables, byres, and bullock house, all form part and parcel of the sides. The harness is never seen about the stable, but is hung on stands in the harness-room, and the corn is kept in iron chests. All the houses have large skylights, and are ventilated by an open ridge ventilator running along the whole length of the roofs, which is further assisted by the introduction of cold air from pipes laid underneath the floors, and communicating with the outside of the buildings. The water service in the cowhouse is furnished with a cock and waste-pipe for each double stall; and each stall in the calf-house, which is quite on a miniature scale, is so managed that the calf cannot turn round, and soil it at both ends. A detached building, lighted at one end by an iron oriel window, is fitted up with iron railed boxes 13 feet by 12, for forty bullocks. There is a raised alley between the rows, but the bullocks were still afield, and we had not the luxury of an "over-sight;" no small gain in showing off any beast, as buyers of bull calves have found to their cost, when the ring-ground has been well chosen.

The dairy stands a few yards off, among some fine old trees, and adjoins the poultry-yard, which has long been famous for its breed of white Dorkings. A larch bole does duty as table in the dairy parlour, and in its polished surface we read its own infallible ring register of some fifty years' growth. The little fox in fire-clay on the chimney-piece has no breathing type about Scone; pheasants scurry in troops across the path as we wend our way through the grounds; and hares get up all round us at Waterloo slip distance, as we cross the hundred-acre field in front of the Palace, and find West Highlanders and crosses, nearly ten-score strong, edging away to the water and the woodlands, to be out of the noon-tide heat.

There is very little breeding in Strathmore, and Falkirk September and October trysts and the Yorkshire calf-men are generally looked to for stores. The Earl's flock of Leicesters numbers about six score ewes, principally of Border blood; and the wedder and ewe hoggs are all fed off the first spring. A thousand blackfaced three-year-old wedders from Cullow are also fed off on turnips, and consigned by the end of May to the Edinburgh market, when they kill up to 20lbs. a quarter. In the October of '63 they cost 36s. 6d., and left from 14s. to 18s. behind them. This is very much the system of management in the Strath, and those who do not depend upon the Cullow wedders buy half-bred and cross-bred lambs at Melrose, and send them fat to the Glasgow and Edinburgh butchers as soon as they have been clipped.

Paton's gun shop is to Perth what Hugh Snowie's is to Inverness—quite an arsenal and lounge for shooters; and evening after evening he sends off nearly 3,000 cartridges to the moors all round. Few men in the provinces make more guns and rifles; and out of his seventy last season, only one was a muzzle-loader, and that was for pigeon shooting. Since Sir A. G. Cumming, of Altyre, brought down five stags out of one herd at six shots, the muzzle-loader has not been able to hold its ground at all. Mr. Paton alters them into breech-loaders, without subjecting them to the injurious process of bringing the barrels to a white heat, and thereby injuring their temper. His plan also does away with the use of the common lever over the guard; and Earl Mansfield, one of the most experienced shots in the North, has been the first to adopt it. Stags' heads are not so much of a Perthshire specialty; but capercailzies, cock and hen on a mossy perch, and a white grouse and starling are Mr. Paton's type of one kind of bird life, and a brown golden eagle and a speckled fishing eagle from the wilds of Athole, of the other. Landing-nets represent the interests of the Tay and its tributaries, and so do fishing-reels, which go round a "desk," instead of a handle, and therefore never get hampered. There is also a drawing of Lord Henry Bentinck, who is probably crawling at that moment on his hands and knees up Glen Fishie, to get within death distance of a hart. The huntsman's horn is silent, and likely to be, as the Perthshire Hunt is a bygone; and in the "vermin" column of the game-book we read of the fox and otter in the distinguished company of not only blue hares and rabbits, but of "ravens, hawks, and magpies, jaypies, and huddy crows.'

Prices are going up fast as the shooting leases fall in. Strathconnan Forest, with the grouse shootings on the estate, has risen from £1,400 to £2,000 in ten years, and two other deer forests, pure and simple, nearly as much. A grouse moor, which was let at £300 two seasons since, has all but doubled itself in price; and another, which twenty years ago could hardly tempt a tenant at a tenner, now brings in its £200. In fact, the letting of moors has become such an important item in a landlord's calculation, that on the recent sale of a Highland estate, after duly taking into consideration the contingencies of non-letting in a bad grouse year, the value of a shooting was capitalized at twenty years' purchase. For some seasons past, the best Perthshire price for dogs has been £28 a brace, and the fashion still runs upon pointers in preference to setters, as they are found to be hardier, and to do much better without water.

When we returned to Perth the next month, there was very little need of placards, bidding us to "Beware of giddy joys which cheat and wound the heart," as none were to be found on the Inch during the celebration of the Caledonian Hunt. The horoscope of the races was very quickly told. *"There are about a dozen horses here—just the regular lot; and they've arranged all the races;"* and there seemed but too much truth in it. Lord Glasgow, Lord Strathmore, and Mr. Stirling Crawfurd are all members of the Club; but they scarcely ever run horses in Scotland; and as Mr. Sharpe had then retired from the turf, there was no one to knock over the arrangements of "The Confederates," or, at all events, to make them race for their money.

Where was Mr. Tom Parr? Was he too busy among the grouse of Kildonan, or the pheasants of Challow? Why was Mr. Merry a saunterer among the English flesh-pots, far away from the sward which he used to frequent in his hot youth with Beardershin and Florentia? Not one Dawson, Tom, Mat, Joe, or John, was to be seen. Robert did not journey to the old spot from Middleham; and John Osborne was, pen in hand, in that little Ashgill parlour, deep in the stars and Weatherby's sheet calendar, and calculating the spoils of another Nursery Stakes.

I'Anson was alone found faithful among the trainers. Nothing can wean him off Scotland; and there he was, snatching every spare minute of dry weather at Perth to have a round at golf — a game at which he is remarkably clever. He was armed with his favourite snuff-box, a silver oyster-shell, presented to him by a backer Caller Ou, and his friends fed their noses copiously. The "land of the mountain and the flood" is the inexhaustible store-house for him on his christening days, and Caller Ou, Blink Bonny, Bonny Bell, Breadalbane, Balrownie, Blair Athole, Broomielaw, Blink Hoolie, and Blooming Heather are the results of the inspiration which he gathers north of the Tweed.

It made one quite sad to look at the meagre card. Where were Lord Wemyss and Philip, with Simmy Templeman in the blue and black cap? Sir James Boswell, General Chasse, and old Sunbeam? Meiklam, with Modesty or Inheritress? General Sharpe, with Canteen, and his brother with old Leda, the foundress of a race of heroes? Ramsay of Barnton, with Lanercost, Inheritor, Despot, or The Doctor, those doughty champions of the yellow and green sleeves? Lord Eglinton, with St. Bennett, St. Martin, or the never-failing Potentate? Fairlie with Zohrab, of the stout heart and the ready kick; or the grey Pyramid, who finished his days in the mail? All gone, and only Joey Jones, Unfashionable Beauty, and Dick Swiveller, beloved of Tom Ruddick, remain.

Two o'clock came, but only a few umbrellas were

going down the street. It had been long known that there would be no race for the Guineas, and that Caller Ou was to manage her thirteenth journey on "the Queen's service" in peace. Things had been arranged to save the old mare the trouble of making an example of the hacks; and even the owner of Unfashionable Beauty wanted to share the spoil. The owners could'nt come to terms, and so the chesnut was saddled and went to the post; but after starting and galloping a short distance, she pulled up, and returned calmly to the enclosure, and the St. Leger mare was vigorously cheered in her canter. The Whip, too, was unchallenged for. The grand stand was but half filled, and only three carriages came. What a change from the time when James Moray, of Abercairney, the first man who established fox-hounds in Perthshire, drove his four-in-hand on the Inch, when he kept the ordinary in a roar by discoursing like an old woman in the soundest Scotch, with a table-napkin round his head, and when he never flagged with heel or jest till the morning sun had peeped into many a ball-room.

There were no drinking-booths, and merely a gymnastic exhibition tent and a wax-work caravan at the very edge of the course. The whole thing wore quite the primitive Newmarket air, before telegraphs were invented, and before a patent betting-ring supplanted the pump on which Pedley was wont to " clear his fine voice, and give a warning thump." Still, if the day had only been fine, the whole scene would have had quite a brown-bread relish, after the formalities of our five great meetings, conducted as they are with as much precision as a Glynn or a Westminster Bank parlour. Two small stands, with striped curtains, sufficed for the Perth and Caledonian Hunt members; but the "bits of pink" were hid by the overcoats and the scarlet cap of a gentleman-rider ready dressed for Bonnington was the only bit of colour visible.

There were the Earl of Mansfield, Lord Stormont, Mr. Whyte Melville, Mr. John Grant, Captain Thomson of The Fife, Mr. Willie Sharpe (hugging his plaid as of yore), and others whom we did not know by sight, all evidently striving to make the best of things. Mr. Nightingale, the great ex-coursing judge, stood on his platform, facing the members' stand (which is not on the same side as the Grand); and the weighing and dressing were conducted in two little canvas tents. Mr. Elliot, that "Manning of the North," was the clerk of the scales and starter, and the members of the Perth Hunt had a comprehensive view of the weighing from their stand above. The telegraph was certainly a weak point; and it was amusing to hear Tom Ruddick rush after Mr. Nightingale into the dressing-tent, and ask what had won, as " there is a lad outside with some bets on"; and so the chief justice duly referred to his pencil entries thereof. The betting was quite a farce. A few speculators walked about, and did a little on the Hunters' Stakes; but there was no one to devour, and they had no chance of preying on each other. Mr. Barber was among them, looking resigned and incredulous; and when the horses did come in, Young Perth rushed wildly up the course with a zest and a yell which did them honour. A race was a race to them, and "blue jacket and black cap" won in a canter. What did they know about arrangements, and all such subtle mysteries? When a race wasn't going on, they could "Try your weight gentle-*men*", or gather round one of the ballad-singers, to whom Fred Turpin was listening with rapt attention as if for the first whimper in a Fifeshire gorse. We should have seen far better fun at Ayr the next year, or at Kelso the year before, but '63 and the Inch were our lot; and the best omen for the Club that day was "a vision of fair women" inside and outside a county drag, all bound for the ball at night.

The Club was instituted at Hamilton in 1777. Her Majesty is the patron; and the eldest living member is the Marquis of Tweeddale, who was admitted in 1809. The uniform is scarlet with green collar, and in old days the slightest variation in the shade of the green would have been spied out in an instant, and two buttons instead of three at the wrist would have formed the subject of a special demurrer. Earls Glasgow, Wemyss, and Moray rank next in seniority, and date from August, 1822; the Duke of Buccleuch is five, and Mr. Little Gilmour six years their junior. It is limited to seventy members "connected with Scotland by birth or property;" and the rule is so rigid, that at present there are forty-one candidates, including a duke, a marquis, three earls, eight lords, an honourable, and six baronets biding their turn. Mr. Sharpe is the only honorary member, in consideration of his services as secretary for more than thirty years to the Club. Mr. Gillespie succeeded him in '62, Mr. Campbell of Blythswood is the present preses, and Col. Mure the treasurer.

The two latter officers and three counsellors (who seem to have no defined duties) are chosen every year, and four form the wine committee, who have to report in writing on the state of the cellar to the December meeting. The entrance-fee is forty guineas, and the annual fee ten; and the ordinary club meetings are held once a month for six months in the year over the mahogany at the Douglas Hotel in Edinburgh, where a saddle of mutton (blackface, of course) is the standard dish. In short, it is one of those grand old-fashioned institutions which struck its tap-root deep in the last century, and, except in its racing, knows no decay.

The Hunt had once the choice of eight places; but Aberdeen, Stirling, Cupar, and Dumfries have gradually disappeared from the roll. There was no regular rotation preserved; but if a neighbourhood wished to have it, the country leaders put on a strong whip at the December meeting of the Club in Edinburgh; and on one occasion the fair canvassers were so active, that forty-five members were mustered to vote.

Once only, in 1823, did the Club go as far as Aberdeen to join the Northern Meeting, when Lords Panmure, Kennedy and Huntly, Sir David Moncrieff, and Sir Alexander Ramsay had studs; but there was a good deal of row, and the Secretary narrowly escaped tossing in a blanket on account of the badness of the wine. The Welter Stakes, 13st. each, for "regular hunters of the preceding season," was the Northern St. Leger; and that year Lord Huntly's Hospitality beat Lord Panmure's renowned Harlequin. The Meeting seemed to die out in 1829; and it was revived in 1843-44, when Zoroaster and The Dog Billy (so called out of compliment to Captain Barclay's gladiator) had a pretty good time of it; but there was no life in it after that.

Cupar Fife was very celebrated for the balls in its Oval County Room; and among those who are dead and gone, it ranked Lord Leven, Sir David Moncrieff, Sir Ralph Anstruther, and Mr. John Dalziel of Lingo, as its greatest supporters. The course was four miles out of the town, and the horses stood at a little country inn called "The Bow of Fife."

Both at Ayr and Perth the ladies come to the race ordinaries; and at Dumfries they were wont to join the public breakfasts. The Perth Hunt, although obsolete in the field sense of the word, has always kept Perth up to its turn; and the late Duke of Athole threw a great deal of life into the balls, with his piper and national dances. Both here and at Dumfries the dancing was much more lively, and there was much less stiffness in every way. The Southern meeting amalgamated most joyously at the latter town, while the Marquis of Queensberry, Sir William and Sir John Heron Maxwell, General Sharpe, and Mr. Alexander of Ballochmyle lived; and once eight-and-thirty horses came, of which Muley Moloch and other cracks walked on from Carlisle. Stirling had a great meeting one year, when the ball was held in the Corn Exchange; but Mr. Ramsay and Mr. Forbes of Callender died soon after, and the thing was given up. Although his Grace cares but little for the sport, the Duke of Roxburghe has manfully supported the meeting at Kelso, where the balls and the sport have been equally good. At Edinburgh the ordinaries failed, and there has never been a year to compare with 1828, when the Duke of Buccleuch was preses, and the town was full to the garrets. Ayr has always had a good county attendance, more especially when the late Lord Eglinton, Sir David Blair, and his brother, Col. Blair, kept horses; and the list was got up in great style, with three £100 Plates in it.

"The Caledonian Hunt" is first mentioned by that title in a race-list, at Dumfries, in 1788, when His Majesty's Plate was granted; but it is enough for our purpose to trace it back for five-and-forty years. In 1820 there were nine meetings in Scotland, and among them the Fife Hunt at Cupar, one at Irvine as well as Ayr, and one at Stranraer, which never recovered the prowess of Anastasius and Archibald, and expired in '22. At *Dumfries*, Charles Lord Lord Queensberry almost cleared the board with Fair Helen, Miss Syntax, and Gonsalvi; and what there was left, Lord Kelburne picked up with Chance, and Sir William Maxwell with Clootie and Monreith. At *Ayr*, next year, Miss Syntax and Fair Helen had matters pretty well to themselves, but in the Springkell colours. At *Edinburgh*, "t'ould grey mare" met, and beat Jock the Laird's brother, and Monreith, both of them winners. Then came the only *Aberdeen* meeting, which began on the Saturday, and went on all the next week. Sir David Moncrieff won eight races, and fairly held the belt with Negotiator; and two out of "the three Yorkshire Tommies"—Lye and Shepherd—were uncommonly busy upon winners. At *Kelso*, in 1824, Sir David had the best of it again in another seven days' bout; but Fair Helen's day was over, and she and Negotiator (now Lord Kennedy's) were both beaten by the lucky baronet's Catton. There was nothing remarkable at "*Ayr*, or Air"; but in 1826 they were racing till dark over the North Inch at *Perth*, and ran two heats and a match next day. A couple of Ardrossan horses, Sir A. Ramsay's Gift and Mr. Baird's Sir Malachi Malagrowther, had the cream of the thing between them, and contrived to keep clear of each other.

The six-year-old Springkell was then great over *Dumfries*, where Lye won the St. Leger on his own horse The Corsair, and Mr. Gilmour a "12 st. each" match on his Minstrel against Sir James Boswell's Boreas. Lord Kelburne's Actæon and Harry Edwards were all the talk at Musselburgh in 1828, when they had won the Gold Cup, and also beaten Queen Elizabeth, Malek, and Springkell. Sir William Maxwell's three-year-old grey Spadassin won the Caledonian Cup for Scotch horses this year, as well as at *Perth* the next, where the stylish, showy Gondolier was in as great force for Lord Elcho as Leda and Conjuror had been at Musselburgh, and as Brunswick was in '30 at *Ayr*. This meeting was peculiarly memorable for the five two-mile heats for the Western Meeting Plate. Five out of the eight did'nt go for the first heat, which was won by Leda. Three did'nt go, two were drawn, and Agitator won the next; then three more were drawn, and Charley, Leda, and Agitator ran a dead heat, and finished in that order for the fourth heat. By this time Agitator had had enough of it, and Charley won the decider. The three were in a four-mile race next day, when Leda and "Sim" reversed matters with Charley, and Agitator was only fifth.

Next year Charley and Leda tried their strength once more at *Kelso* in two two-mile heats, in which the mare was worsted. The horse was never in greater form, and so Ballochmyle found in the Caledonian Plate, and The Earl in The Guineas. Lord Elcho won a match on Brunswick against Sir David Baird's Queen Elizabeth, while Nicholson won for his lordship on Gondolier, and "Sim" on Leda once more. Three four-mile heats against Philip, Terror, and Round Robin killed Ballochmyle at Edinburgh, or rather at *Gullane*, in '32, where the races had been removed on account of the cholera; and at *Edinburgh*, the next year, Lord Eglinton had the best of it, both in a match against Mr. Gilmour and the Caledonian Coplow; and "Sim," in the Duke of Buccleuch's colours, twice finished in front. General Chassé, with his two wins and his two walks over, was quite the Platoff of the *Dumfries* meeting; and Inheritor, whom he beat for the St. Leger, made short work of Muley Moloch and Philip at four miles for the Guineas. The black and the chesnut's places were reversed at *Ayr*, next year, in the Gold Cup; and in the Caledonian Cup, the defeat of Sir James's horse by Myrrha was quite "a surprise." This was a year of promise to Mr. Ramsay; and at *Perth*, in '36, Mr. Merry made a somewhat unpromising opening with Florentia, to whom the evergreen Philip (who had just won all before him at Lamberton Moor, a sort of rival to Gretna in its day) showed no mercy. Inheritor was out of form at *Musselburgh* in '37, and the star of Mr. Meiklam and Modesty began to rise, and Mr. Wilkins's breeding to tell both with her and Abraham Newland. The Merry yellow jacket, which was destined for greater triumphs than them all, made a better appearance at *Ayr* in '38, where Inheritor was himself once more, and Lanercost convinced St. Martin twice over that he was "no use to him."

This Cumberland brown, which Mr. Ramsay bought from Mr. Ramshay* for 1,500 gs., had now the championship of the Scottish Turf, and held it against Bellona, Malvolio, and all comers, at *Cupar Fife*. Then came that memorable *Kelso* struggle, in which, with The Doctor to help, and under very high weights, he beat Beeswing for the Gold Cup, cleverly and with nothing to force the running, and 2lbs. the worst of the weights, which had been reduced about a stone, finished level with her in a two-mile plate that same afternoon.

At *Stirling* it was the turn of Mr. Meiklam and the dark blue and white stripes with Broadwath and Wee Willie; and Charles XII. and Job Marson were at *Perth* the following year, to look after The Whip for Mr. Andrew Johnstone. At *Ayr*, in '43, it was nothing but "dancing after a *shadow*"; but that mare had to yield in her turn when Alice Hawthorn met her at *Dumfries* next year. In '45, Pilot led the way at *Kelso*; and Mr. Merry's prowess at *Perth*, Lord Eglinton's with Eryx and Plaudit at *Ayr*, Inheritress's and Chanticleer's at *Edinburgh*, and Elthiron's at *Stirling*, bring matters up to '49. Elthiron ran the wrong side of a post next year at *Perth*, where Haricot had quite a blaze of triumph. Old Clothworker tried his hand at *Ayr* in the Exhibition year; but although his owner was on him three times, at 8st. 10lbs. and upwards, the chesnut was

* This will correct a slight typographical confusion between these names in "Scottish Racing."—"SCOTT AND SEBRIGHT," pp. 180-192.

beaten five times in all. Miss Ann was the bay queen of the revels at *Edinburgh* in '52 ; and then *Kelso* had one of its crack meetings, in which Balrownie, Goorkah, and Defiance divided the honours. *Stirling*, in '54, was quite a Wild Huntsman carnival; and Lord Eglinton's colours caught Mr. Nightingale's eye for the last time in the handicap with Bianca. Braxy, Braxy and Heir of Linne, and Sprig of Shillelah had the lion's share in 1855-57 at *Perth, Ayr,* and *Edinburgh.*

Ignoramus was the crack at *Kelso;* and Underhand, after having everything his own way at *Perth* in '59, was beaten twice by Caller Ou at *Ayr* in '60, and was last to Caller Ou at *Edinburgh* in '61. Fobert had, however, an avenger of his "little bay horse" in Oldminster, who beat the St. Leger mare at 5lbs. in a canter, and at evens by a head, with Peggy Taft to look on each time. And so "the ball was kept rolling" at *Kelso* in '62, when Haddington was all potent; at *Perth,* of which we have told before; and finally at *Ayr,* where St. Alexis, a cheap Doncaster purchase, did just what he chose with his fields, and Joe Graham, the little Dumfriesshire huntsman, achieved the great *coup* of his life by winning with his £50 horse Blood Royal that Caledonian Cup for which the highest families in Scotland have often longed in vain.

(N)

CHAPTER XIII.
DUNKELD TO BLAIR ATHOLE.

"I drove up to Loch Ordrie, and home by Craig-y-Barns, fifteen miles through woods of my own planting, from a year to forty years old, a very grand picturesque drive, not to be equalled in Great Britain."

The Athole Larches—The late Duke of Athole—His Love of Otter Hunting—The Athole Forest—The Dunkeld Ayrshires—Milk Statistics—The Duchess Dowager's Farm at St. Colme's—Pitlochry—The Castle and Burial-place at Blair Athole—The West Highland Herd.

So wrote John Fourth Duke of Athole, the most energetic of Scottish planters. Mr. Christopher Sykes, of Sledmere, who received a Book on Birds from an East Riding Society in 1780 for planting 54,430 larches, was as nothing to his Grace. In larch, and larch only, "with its sharp-pointed top, which gives no rest to the snow," he placed his woodland trust. Six thousand five hundred acres were planted with it solely, and he gave it seventy-two years to build a navy. Each tree was to produce fifty cubic feet, and they were to sell for £1,000 an acre, plus £7 an acre for the thinnings. Everything conspired to strengthen his conviction. The Navy Office reported the "Atholl" to be far sounder than the "Niemen" frigate of Baltic fir. The larch was found not to splinter so easily, and to fire more sluggishly when it was tested by a broadside and a brazier; and when "The Larch" was finally wrecked at Ladra on the shores of the Black Sea, it fought with the breakers to the last. Such were the hopes of the Duke when iron-clads were a vision only worthy of "a horse marine," and the larch had not learnt to die downwards.

Disease may have done its work among them, and thousands fell one February night when the storm swept down Strath Braan; but there they stand, thickly crowning the summit of Craig Vinean and Craig-y-Barns. These rocky outposts of the Athole woodlands are seen in all their grandeur from the Abbey grounds. The silver fir and the bonnet fir, with its finer bark and smaller thread, flourish in these smooth over-arching glades by the side of the Tay; and there, too, are the very Gog and Magog of the larch clan, a hundred feet each in stature, and four hundred and eighteen feet cubic in bulk. Tea leaves are said to have been cooked cabbage-fashion when they first arrived in England; and tradition has it that these larches were nursed in a hot-house till they pined, and then took root by chance, and flourished when they had been expelled from it as worthless. Birnam Hill, with its purple heather circlet, stands proudly in the distance; but its "wood" is reduced at last to an oak tree and a plane. The sun lights up the speared otter on the kennel vane; but Cruel and Conqueror have been gone these seven years. Only a few black-horned and half-bred St. Kilda ewes are browsing on the slopes; and the prize Ayrshire quey at Stirling, and the one that beat her at Glasgow, are in the field below.

Otter-hunting was at one time quite a passion with the late Duke, and many a long day he had with that kindred spirit, Lord John Scott, on the Teviot and the Tweed. He was a man of immense muscular energy, and forty-four inches across the chest. His pace at starting was not great; but, solaced by his unfailing pipe, he could stay for ever. One peculiarity was greatly in his favour. When he was coming home, he would stop every third or fourth mile, and sit down by the roadside, fall asleep in an instant, and at the end of five minutes start on to his legs again, and off like a new man. Beardy Willie (whose "tumbling cataract of beard" fairly appalled us as he opened the gate at Dunkeld) not only acted as his banner-bearer when, as Knight of the Gael,

"He went
Gaily to the tournament,"

but was always a henchman on these occasions, carrying a spade and a pickaxe on his shoulder from the dawn to the gloamin, and as full of running as his master, from end to end.

The Duke would hardly eat or drink anything when he was with his pack. Once he started early in the morning, and hunted the Braan for ten miles, then back for a couple, and crossed the hill from Amulvie to Aberfeldy, to try two lochs. After that he hunted part of the way back to Dunkeld, and got there at eleven o'clock, very little the worse for fifty miles and a blank day. About twelve couple was his favourite number; but once, when an otter had wearied them out below the Stenton rocks, he sent back for seventeen couple more. His first pack was made up from the Marquis of Worcester's and Captain Hopwood's; and he delighted to tell how Manager lay six hours in a drain below Kinnaird, and only spoke when he went in; how Jesuit never left the water for six or seven miles on The Tweed; and how land and water seemed alike to London and his celebrated Carlisle pair Conqueror and Cruel. The season began in April, and he hunted two days a week for four or five months on seven or eight rivers. His maiden otter was killed on the Braan, but he never bore home a single trophy from the Tilt or the Garry.

He once had harriers at Strathord in Perthshire, and would occasionally handle a deer with them, and ride, despite his short-sightedness, with an energy that was almost miraculous. While in the Scots Greys, he had been a hard rider, and Ben-y-Ghlo and Confusion, which won the Hunters' Stake at Perth, were his best-known thorough-breds.

If there was one finer reminiscence than another of his indomitable pluck, it was when he rode his Eagle for the Perth Hurdle Race in '38. He fell at the second hurdle opposite Marshall's Place; but though his collar-bone was broken, he would be lifted on again, and caught his horses and won. The Erl King could'nt have ridden faster if he chanced to be a minute late in starting for the Drive; and a comical sketch of the kind hung in the drawing-room at Blair Castle from the pencil of his Eton friend, Mr. Evans. Grouse-shooting he did not care much about; but the autumn drives, of which he would have at least a score, if the wind suited, were his delight. Glen Tilt will average about ten thousand deer daily, and there is no finer sight than an army of harts moving along its sky-line. The great Athole Forest comprises some 80,000 acres, and marches with Glen Fishie, Lairg, and Mar. It has been known to carry 30,000 deer. If the wind is from the south, it is the best drive for Athole into the heart of the Forest; and if in the north, it is the best for Ben-y-Ghlo. But we know nothing of such forest lore, and the wind rancorously barred our only chances of seeing a charge of six thousand down the glen.

Still, if this was denied us, during the few autumn days we lingered near Blair Athole, it was no light thing for an Englishman to have noted the allegiance of Highlanders to a chief, so deep and true in his days of health, and deepened and chastened when he was dying. Not till then could we have realized the truth of what was written, "that his Athole guard (many of whom, with Struan at their head, were his peers in birth) would have died for him, not in word, but in deed"; and that "a young capable shepherd, who might have pushed his fortune anywhere and to any length, was more than rewarded for living a solitary deer-keeper at the far end of Glen Tilt, or some nameless wild—where for months he saw no living thing but his dog and the deer, the eagles and the hill-fox, the raven and the curlew—by his £18 a year, his £3 for milk, his six bolls and a-half of oatmeal, with his annual coat of grey tweed, his kilt, and his hose, so that he had the chance of a kind word or a nod from the Duke, or, more blessed still, a friendly pipe with him in his hut, and a confidential chat on the interests of the ' Forest.' "*

It was something heroically grand to see his Grace in the middle of the ring at Kelso, all muffled up, to hide the ravages of that terrible malady which was so surely eating away his life, and yet calmly giving orders about his cattle, Ayrshire and West Highland, which "*will not be scattered, when I die, to the four winds of heaven.*" The servants of the exhibitors had not forgotten how unceasingly he watched over their interests at Battersea; and they seemed to vie with each other in holding his horse, and anticipating every wish. His death, as he told his friends, when he bade them good-bye in the yard, with as much calmness as if he were only taking a journey, "*may be a matter of only ten days or three months*"; but it was six weary months before he found rest. When he could hardly speak, we saw him leaving the Dunkeld station to superintend the

* *Scotsman.*

trucking of some Ayrshire cows, which he had selected for the royal dairy at Balmoral; and later still, when Her Majesty paid her visit of sympathy to Blair Athole, no pain or weakness could restrain him from accompanying her to the station. Then for the last time his Highlanders heard his voice, as, after kneeling to kiss Her Majesty's hand, he strove to dispel the gloom of that sad parting by giving them the word for three lusty cheers.

It was as president of the Highland Society in 1858-61 that he first took a fancy for farming and Ayrshires, of which her Grace had a few at Dunkeld. A certain number of them always accompanied him to Blair Athole, and he took them with him the first time he ever went there by rail. He left Dunkeld for Blair Athole some months before he died; but to the last he had a weekly report of the milking and specimens of beetroot, mangel, and kohl-rabi sent to him from his farms at New Tigle and Haugh End.

The dairy books were kept during his Grace's lifetime with the most scrupulous exactness; and Her Grace the Duchess Dowager, and her friend Miss Murray MacGregor have taken an equal interest in the herd since his decease. The Dunkeld Herd-Book is illustrated with notes on the marks and properties of nearly every cow. Attiquin was the first purchase, but she earned no special mention. The red Premium, "with head and horns very small," is noted as a first-prize taker at Glasgow, and as being taken over from her Grace's stock at Dunkeld along with Frew, "not a good milker and always inclined to fat," and twenty others, of which thirteen were not carried on. Carrick, "a very glossy white with shaded tan-spots," is recorded as having given eighteen quarts a day at her best; Cherry for having "one horn growing down and being somewhat wild;" Langholme for "her rather short neck"; Gaiety for "her very gay head and horns erect"; Empress as "very small and beautifully made"; and Bertha as "red, very thin, with a venerable appearance."

Glengall has the welcome entry of "twin queys" against her name, while two bulls and a dead quey are credited to Colly Hill. This queen of the Ayrshires was bred by Mr. John Craig, of Colly Hill, Strathaven; and her description is "white with rich red spots beautifully marked, very large body and short fore-legs."

The herd never exceeded fifty cows in milk, and including a few cross-breds and a Jersey, her Grace has nearly as many at present, and about one hundred and thirty head in all at Dunkeld. The milk was first reduced to the present careful system of calculation on March 31st, 1859; and on Thursday every cow's milk is weighed separately, and multiplied by seven, to get the amount for the week. If cows are away—as the two Premiums, Colly Hill, White Legs, and Idiot were, at Kelso—an average is taken from the preceding and following week, so that it is possible to look back over the books for more than six years, and approximate within a fraction of what every cow has been doing. Taking the average of the largest week in the height of the grass, we find that 702 milkings produced 8,100lbs. of milk, or 810 gallons, from which there came 39½ gallons of cream. That Ayrshires differ very widely, may be seen by comparing this nine-quarts-a-day average with Glengall's, who averaged 12 quarts and a fraction daily for fifty-eight successive weeks. For seven weeks, between May 7th and July 3rd, she averaged a trifle beyond 6 gallons a day; and Colly Hill, who led in 1860-61, still gives 4 gallons at her best, and beat them all save Marion last September.

They generally begin with cooked food in November; and a two-horse power engine, with a corn grinder, oil-cake bruiser, turnip pulper, and hay, chaff, and straw-bedding cutters attached, is fixed in the boiling room at the end of the byre. The four boilers are filled with hay cut two inches long, rapecake, and bean meal in layers, and then steamed; and the large waggon, which runs on a tram-road with turntables, bears two pails of mixture to every cow per day. The four milkmaids have each a soap-box, a towel, and a currycomb. After each milking they scrupulously wash their hands, and they keep their pets in winter as bright in their coats as a blood horse. At one time it was the regular milkmaid fashion to shift sides so as to balance the vessel, but it was found to do no practical good, and the cow often became shifty and kicked over the pail. Each rake or 30lbs. of milk is weighed, and then carried to the tin dishes in the dairy. The calculation is that one gallon should be equal to a pint of cream, if it is fine weather and it rises properly, and that a quart of cream should produce about a pound of butter; but this is hardly borne out in practice. Only skim-milk cheese is made, and nearly a hundred flagons at the door of the dairy were ready to receive the milk of the night before, and to disperse it at a penny a quart through Dunkeld.

We sat down with Mr. Christie, who has always taken nearly as much interest in the Ayrshires as his late master, to watch the "kye come hame" through the Arnagag pasture. The bell-cow had renounced her privileges for the day, and Maxwell, a present from the Secretary of the Highland Society, led the van. Kilsyth, one of the prettiest, succeeded, and then Whitelegs, with a head and neck as sweet in their way as Atty's vessel, which is, after all, the point on which the Ayrshire judge's eye first concentrates itself. Hopeful, for instance, is a nice cow, but then "her vessel is gone for shows." Premium, a gold-medal cow of great depth, came next with the bell, and then Marion, the milk belle of the herd, and good for nine quarts at a meal. When Glengall was in her prime, she would sometimes not stop short of thirteen, with only a few peas and a little bean meal to aid her. She was bred by Mr. Wallace of Kirklandholm near Ayr, and was purchased, after winning at his Grace's milking competition, at a pound for every pound of the highest day's produce. She is "no in tid" now, and she passed on with The Quey, Jersey, and Risk, who has a deal of Colly Hill about her. Then we note the deer-horn and speckled white flank of the red Empress, a prize taker at Ayr, the rich colour, fine muzzles, and rare vessels of Strathaven and Brocky, a first and second at Perth, Maybole and Dalrymple of the old style, Idiot who will keep her head in the manger listlessly for an hour together, Queen of Hearts, and the loving, speckled pair Bryony and Susan who are never ten yards apart. Coda, with the fine deer-like eye, has a beautiful vessel and a good enough tail, albeit a tale hung thereon at Ayr; but, alas! Belle's teats do not hang perfectly square.

Colly Hill is a living proof that a really orthodox vessel will stand after six. Her early promise was not very remarkable, as she was sold as a two-year-old for £14, and gradually rose from £20, £40, and £80 to £90, at which price his Grace bought her after she won at Edinburgh in 1859, and then refused £400 in two quarters for her at Battersea. A portrait of her by Gourlay Steell, with the head dairymaid at her side, is over the sideboard in the dining-room at Dunkeld. There are no queys from her as yet, but she had a nice, little, younger son, in the calf-house; though a climb and a peep at her oldest son, Colly, in the shed near the saw-pit, was not so satisfactory.

From Dunkeld we proceeded onwards to St. Colme's, a farm which her Grace has held for some years, and fitted up one wing for an occasional residence. The Tay and the Tummell meet in the valley below, and stretch away to Pitlochry and the Pass of Killiecrankie, while Ben-y-ghlo rules in the mountain background. About eight hundred acres of arable and grazing land are attached to it; and a great many beasts, of which there is a large annual sale, wander away among wood and fell, and are collected and counted once a day. Dorkings, cinnamon turkeys, and Aylesbury ducks collect in hosts at feeding time in the yard. Mrs. Kennedy (the wife of the farm manager, who was then very ill, and never rallied again) was still more proud of the polished pine wainscotings, and the beautiful order of the granaries, where every grain seems to fall into its place. There is a combined thrashing and winnowing machine, with a travelling band to raise the grain; and a view from the back of the granary suddenly revealed an Ayrshire "by a gold-medal bull, if he's not one himself." A sale of "Spare Ayrshire Stock," which is probably the germ of an annual one, was held here last March, and the six young queys averaged £16 3s. 9d.

The train only ran as far as Pitlochry, the first year that we were at Blair Athole, and we had an opportunity of attending the fair there. "Parlies were not up" that year, as they were on the one following, and the proceedings were anything but animated. Nine or ten men were hanging over some little blackfaces, but, with the exception of a remark that one was "*no a despisable beast*," and that the whole were "*six quarters old*," there was nothing to break the flood of Gaelic disputation.

Blair Athole is truly the land of the kilt and the claymore, and no one in the late Duke's employ dared to shirk his nationality. Everything is in keeping with an old Highland home—that Blair Castle to which "Castle Rushen sendeth greeting" in rude letters on a horn. Its Long Passage is decked on each side with the skeletons of antlered heads; and among the pictures are Neil Gow with his violin, and the muster in '42 of the loyal clan Athole behind the Duke (then Lord Glenlyon) on the green, ere they marched to receive the Queen at Dunkeld. The gallows is still there, a stern relic of the days when a chieftain held the power of life and death. The long ivy tendrils were working through the ruined windows of the old church, where Claverhouse sleeps; and there, too, among those boundless forests which he loved so well and the hills which stand fast for ever, the Athole Guard laid the man of their heart.

The West Highland herd are all kept here; but it was only within a year of his death that the late Duke began to care for it. The remnants of the original herd were sold to Mr. Hendrie of Ayr in the autumn of '63, a few months after the Breadalbane sale. It was there that his Grace bid so determinedly against the late Duke of Hamilton, and bought a bull at £136, three bullocks at £128, and four two and three year old queys for £219, and at Kelso he added some more "Breadalbanes" from Hamilton to his stock. We found Donald, who had not belied his price in the Stirling and Newcastle lists, wandering leisurely with the brindled Oscar and the dun bull in front of a German artist, who was not falling back on the Infinite for his conceptions, but soberly sketching under a huge umbrella from plain beef and blood. The present Duke keeps on the West Highland cattle, and has thirty in the breeding herd, about half of which are cows.

The calves are generally dropped in April, and weaned in October, and run with their dams at the end of three days. Four "Breadalbane" bullocks were at the steading beyond Toll Damh. The light yellow and the dun we had met before, when they reversed their places at Kelso and Stirling; and a light red, with the beautiful fan ear, and a dark red, with a coil of hay rope tossed by chance over his fore ribs as if to challenge the tape-line, made up the shaggy group. They were all four-year-olds; and though Macdonald allows that the "Stirling dun is a braw stot, and the Kelso yellow has not an inch to mend, and horns as straight as you can set them;" still he carries in his eye an unbeaten yellow of Mr. Campbell's of Monzie, which won at Perth and Edinburgh ten or twelve years ago, and considers him far ahead of them. They were living on the best of every thing for Christmas; and locust beans, oilcake, and ground corn were among their side-dishes.

From them we wend our way, and bless Duke John for the shade of his larches, down towards Glen Fender or the Black Knoll, the pass between Fender and Glentilt. There is land of every kind, heather and pasture, in that famous Glen; and blackfaces hold all the east side, where it is green, from the top to the bottom. The cows were grazing opposite Black Knoll, whose sides are dotted with roans and whites of two crosses at least, from the Muir of Ord or Amulvie. Below is a dell of ashes, interspersed with the sallow green copses of mountain birch; but, as our guide tells us, "the worm with the red head works away on the leaves, and does them fair out."

An old doddy looks on apart, while the red Newcastle Emily is driven up from the brae. The dun Proisaig Odhar, which was second at Stirling, can hardly boast of such a black bull-calf as that which has "sair suckit" Breadalbane Queen, the first in her class at Kelso. A broken horn marks and gives the name to a fine quartered brindle, and the silver sheen or white dun claims close kindred with Poltalloch's prize Stirling cow Newrack. A dun with a blind eye is not one of the worst, as she draws up to Rosie, the first at Kelso and gold medalist at Stirling, but with this one drawback, that her young Donald is "*a wee thing white.*" Then there is a "Mairi Dhubh," and a "Molachag" or sister to Molachag, and dire verbal difficulties crowd on us. Macdonald's are not merely verbal; and when we remember how the Breadalbane heifers (rather a wild breed to begin with) were hurdled off there, and how nearly we had our own cheek-bone broken in the North, we can quite sympathize with his reminiscences: *"We had a yaie job with the ladies at the Stirling Show—terribly wicked, fearfully wicked."*

CHAPTER XIV.
PERTH TO KEIR.

"Study their monuments, their gravestones, their epitaphs, on the spots where they lie: study, if possible, the scenes of the events, their aspect, their architecture, their geography; the tradition which has survived the history; the legend which has survived the tradition; the mountain, the stream, the shapeless stone, which has survived even history and tradition and legend."

DEAN STANLEY'S EASTERN CHURCH.

The Strathallan Castle Herd—Walk to Rob Roy's Grave—Fern Thatching—The Braes of Balquhidder—Points of West Highland Cattle—Black-faced Sheep—"M'Claren's Cow"—Recollections of Deanstown—Keir—The Steading—Clydesdales, Shorthorns, and Garden.

PLEASANT Perth was left at last. We wended our way through the country once hunted by James Moray of Abercairney, and then by John Grant of Kilgraston, but now unhappily devoted to wire fences instead of foxes, and so on past Auchterarder, whose humble steeple (with what looks like a golden bird pluming its wings to fly, but proves on closer inspection to be a griffin), marks where the Kirk of Scotland was rent in twain. Strathallan Castle and its shorthorns were our goal that day. His lordship began from Bates (whose Wild Eyes did no good), Robertson, and Stirling, uncle of the present member for Perthshire, and bought his bulls from Yorkshire. The principal tribes are from Julia by Belted Will, descended from Mr. Shaftoe of Whitworth's stock, and White

Duchess, bred at Keir, and bought by Mr. John Gardiner of North Kinkell, an old tenant of the Strathallan family, and very fond of shorthorns.

The tenants keep a number of cows, generally of the original Perthshire breed, which they put to shorthorn bulls, either their own or his lordship's, who gives them special facilities at the Home Farm. The calves thus obtained, together with others bought in the district, and occasionally a few stirks from the Falkirks, are reared for stores, which are eventually turned off principally to the Glasgow and Edinburgh markets, at from two and a-half to three years old. The strath of Earn consists more of arable than grass, and is principally on the old red sandstone. Lord Strathallan has about five hundred acres in his own hands, including the Park, which is principally stocked with Norlands or small-sized West Highlanders, which the Northern dealers collect, and drop by scores on the road to their regular customers. They will work up any roughage, and it is not for them that his lordship cultivates his favourite green crop of St. John's Day Rye. He began with it fifteen years since, and generally sows it in June. It is seldom cut before May and then the field is ploughed down, and a green crop is taken.

There is a Leicester flock of fifty ewes at the Castle, of which the tups are sold for a cross to the Ochil Hill farmers, who keep up their Cheviot and blackface ewe stock by buying in lambs; and the half-bred and cross-bred produce go to farmers in the valley, who can now hardly get enough of them. Five or six years ago there was hardly a sheep in the strath; and when an old man got knocked down by a bouncing Cheviot, he said, in the bitterness of his heart, they were "*noe belter than swine.*" Against the latter there is no prejudice, and his lordship's Fisher Hobbses are bought up as fast as they can be bred.

The early history of Lord Strathallan's herd is virtually to be found in one of his photograph-books, which contains his choicest bulls, cows, and heifers, at all ages and in all attitudes. Colonel Towneley's Barnaby Rudge is there, with that deep brisket and rather staring colour, which he brought into the herd; Dick, with the twisted horn, the high tail, the long quarters, and the out-shoulder; the more perfect Retribution, son of Barnaby Rudge; the sweet white Hautboy, from Abraham Parker's dam; Harpsichord, another white; The Squire, bred by Douglas; Red Gauntlet, by Sir James the Rose, from one of The Squire's daughters; and so on to Fosco, the double second of '64 at the Royal and the Highland, and Allan, son of his conqueror Forth. Old Frolic has not been forgotten; Julia, Wild Eyes, and Warlabina have all been "in position," and so have Cobweb with her promising calf, and Rosa Bonheur and Ruby of the drooping horns. There, too, is his lordship's old bailiff, James Thompson, who has always dearly loved a shorthorn, and fed his fancy to the full, when he went down special a few years since to look after the very difficult drainage of Towneley Park.

He was waiting for us on the rustic bridge at the edge of the pheasant woods, through which you reach the steading from the Castle. Some blue Spanish, a breed which his lordship especially values for winter laying, were sunning themselves, and the photographer was taking advantage of the gleam, to hit off Rob Roy, son of Bridegroom. The sweet-looking, massive Fosco, who has not misspent his hours since Stirling, and was walking 22½ cwt., had just been hit off; but white bulls never seem to come out so well as roan, and look too much like the Charolais. Allan, of the thick, level form and beautiful fore-leg occupied the next box, and bids fair, although he rather lacks liberty in his fore-quarter, to make himself felt in the show field.

The red Towneley Violante was in the next box; and then came Rosa Bonheur, who has been placed second and third at the Highland already, Rosebud, Rosa Lee, whose day seemed almost over, the useful roan Warlabina, Fair Maid of Perth, of sound, old-fashioned Ladykirk look and blood, Princess Royal, and Jeanette, a little red, who was third at Kelso, and, save Rosebud, the only Dick left. There were thirty-eight in all, and the calves principally by Allan, one of them as wealthy as himself, but throwing back to the staring, flecked coat of Barnaby Rudge through his dam. Still, the pair of West Highland calves which had come very kindly to hand were quite the little pets of the place; and two or three score of black Essex pigs, old and young, were out at work among the potatoes, in those long hurdled yards which radiate from the sties. Another look at Fosco and Allan, and then we rode on once more to Keir, and, leaving our mare in good box-quarters for a fortnight among the Peggies and the Jessies, we made a dash by rail and foot for a day in the Western Highlands.

The tourist time was over; but, nothing daunted by the snow-cap on Ben Ledi, we left Callander, stick in hand, and, scorning the turn to the Trossachs, we pointed away towards Rob Roy's grave. There was little to see at starting, save the light-brown oak and underwood of the Pass of Leny, where the Scotch fir seemed fairly driven off the ground to an islet. On our left for miles were two lazy lochs, Lubnaig and Voil, joined by a lazier stream. Nothing was sailing on them save a couple of swans; but there is no swan hopping, no municipal marking of cygnets there. The ash had lost its leaf, and the birk, rich with materials for many a bobbin frame, grew by its side; and goats, black, white, and grey, were just visible on the face of a large rock, where it is hardly safe for the sheep to climb. The farms along the road have generally a share of the low ground and of the hill behind. They vary in size from a hundred to a thousand acres, and carry flocks of blackfaces in proportion.

Valley operations were rather at a standstill,

owing to the rain, and in fact many of the cottars are under water a good part of their time. Some of them were on their housetops deep in the mysteries of fern thatching. The best ferns for the purpose are those with a slender feeble stem from eighteen to twenty-four inches, which are hardly able to support their own weight without leaning. From June up to the middle of August they are too soft and spongy for use; and about "Holyrood Day," when they have withered under a September sky from red to yellow, they are pulled up by the roots, bound in sheaves, and stacked, and then laid in straight and regular rows on the roofs, and tightly bound with osiers. The layers require renewing every fourth year; and the fern also does good service as bedding for the cattle and cover for the turnip heaps.

A belted and booted police sergeant stalked along the valley, presenting rather a suggestive comment on the days when the Queen's writ would have run in vain to the lawless braes of Balquhidder. There are tracks in plenty, not of the "hot trod," but of Highland beeves and sheep, along this great North road, which is wakened up summer and autumn by the drover's cry. The Skye and Lewis droves all come this way by Badenoch, Lochaber, Glen Dochart, and Tynd Drum, to the Balgair and Falkirk trysts; but our road turns off at their wonted resting-place, King's House. We still work on by the side of Loch Voil, among wire fences and Ayrshire dairies; and when we asked about the cheese, we only heard that they "made Cheddar, and tried to make Stilton." A chapel with a graveyard on a dark fir knoll gave birth to another query, and we were told in still darker speech: "*It's no Rob Roy, it's Mac Gregor that's stopping there.*" Farther on there is the ivied shell of an old church, which still denies its successor the seisin of the bell; while the forthcoming meeting of the parochial board is nailed on an ash hard by. One stone has no symbol save a horse-shoe; and a half-broken slab, with a broad sword rudely carved, marks the grave of one whom we must now try to regard in the tenderer light of an eminent cattle dealer. There are historic doubts as to whether Rob's deeds were equal to his fame; but his voice, at all events, as it echoed across Loch Voil, rivalled in its volume that Greenwich innkeeper's, to whom Richardson the showman left £1,000 " because he was such a bould speaker," and " could be heerd all over the fear." We still see on the other side of the loch the thatched cottage in which his mother died. That strange chasm in the rock summit still seems to open and shut as it did beneath his gaze; but many a freezing east wind has blown since then, and the heather and the bleaberries have all gone from the braes of Balquhidder. A few deer pass with the storm from the Black and Glenartney Forests, but the blackfaces have the hills pretty nearly to themselves, and the brindled Duncan wanders by the side of the loch "just as canny as a horse," with a dozen of his calves, and cows and heifers in a troop, among the beech, fern, and hazel of Monachoyle Moor.

The West Highland calves generally go with their dam six months, and are weaned in the beginning of October. Their owners prefer them to calve in the house, as the calves always bear the hand better. In fact, the calf is so tameless by nature, that even if it is dropped outside, and brought in at the lapse of a day or two, it is for a long time far more difficult to do with. Nearly every cow has a name and knows it, and it was curious to note at the Breadalbane sale how one stopped its wild capers in the ring when the dairymaid was summoned, and adjured it in Gaelic, as "*Black Precious,*" to desist. The West Highlander's coat is not in its height till October, and hence bulls which you have met in all their glen glory look very different, not to say dejected, in the Highland Society's lists. They require a purple heather back-ground and a leaden October sky. Many breeders dislike too much curl in the coat, and more especially bunchiness at the tail-head. The lighter-coloured ones are sometimes wilder, and the "Argyle black" are generally thought the best and the hardiest, and fetch the highest price. Still, forty years ago the celebrated Dunrobin breed were chiefly brindled; and Lord Breadalbane hung to duns, and bought a bull of that colour, a few years before his death, from Alick Macdonald of Balchillar.

Red and dun are still the favourite colours for stocking the English policies, and therefore they will always be cultivated. The difference {between the tastes of the two countries comes out not only in the colour but the bone, which the Englishman likes to get as light as possible, from the belief that it indicates feeding to a higher weight. Scotland, on the contrary, likes bone and hair to the hoof, and has no sympathy with "fine offal." A short head with a broad forehand is what the breeders aim at; and the horn must be "sappy," yellow to the root, of a fine waxy texture, and with blue tips. Good grass of course tells on the goodness of the horn, and "hard grass hard horn" is quite a settled belief. The horn grows most at two years old, and if the queys are put to the bull at that age, its growth is stopped, and the handling is spoilt as well. Two-year-old heifers will make from £8 10s. to £13 at Falkirk, and bulls range at all prices from £20 to £80. Off grass the Highland cows ought to kill at 50st. of 14lbs., but the Breadalbanes would average nearly 10st. more. There are seldom more than forty cows in a herd, and no owner has ever equalled M'Neill of Colinsay, who had once, it is said, two hundred calves, and cows still choicer than his bulls. Red Water is their greatest enemy, and they are especially subject to it in a cold, wet May.

A shorthorn bull was once introduced by the side of Loch Voil, but the climate was too cold; and even the Cheviots cannot thrive here. Highlanders, like their neighbours, have gone in more for sheep than cattle for ten or twelve years past. The rents of ewe and wedder farmers have been increased; and the flocks, of which Williamson of Glenlochy and Halliday (who marches with the Black Forest) and his neighbour James Menzies have large ones, are nearly all blackfaced, with a very slight Cheviot sprinkling.

Once the sprittle or speckled faced were more the fashion; but now the Irish purchasers like them darker in the face and greyer in the legs. The Aberdeenshire people also look out for the sprittlefaced—not the pure white with spots on the face—when they go to West Linton fair. They always think them more growthy sheep, and when they have not hard but good "rotten horns" (or open at the end), it is generally symptomatic of much better thriving. The West Linton wedder lambs are generally allowed to go about two weeks longer entire than the North or West Highland stocks, which gives strength to the horn and bone. They are, in fact, a stronger class of sheep, and fully better woolled. Many of the tups used both in Perthshire, Argyleshire, and Inverness-shire, are bred in Ayrshire as well as Lanarkshire, and a large proportion of the ewe hoggs are from the latter county as well. The ewe and wedder hoggs nearly always come down to winter on the grass; some go as far as Greenhill and Airdrie, and unless they are highly wintered, they will never realize the five fleeces to the 24lb. stone.

Mr. Lucas of the Bridge of Allan is the largest dealer in these parts, and sends carcases to London, and wedders, of which he keeps many hundreds on turnips all the winter principally in the neighbourhood of Perth, to the Glasgow and Edinburgh markets. He buys most of his fat stock in Fifeshire, but deals largely with the Forfarshire, Perthshire, and Stirlingshire men as well. The largest sheep-farmers in Argyleshire are said to be John M'Kay of Succoth and Martin of Loctraig; and the Perthshire stocks of the Richmonds of Balhaldie and Dron, Elliot of Laighwood, and White of Glen Prossan will run from eight to ten thousand. The blackfaced stocks throughout the two counties range from one to four thousand. White of Glen Prossan and Mrs. Kennedy of Glenmaye (on the Grampians) have perhaps the best blackfaced wedders, which are simply Lanark lambs kept for three years. The cast ewes are generally sold at home, and many of them go as "grit ewes" to the House of Muir market, and are bought up by those who have grass parks. Both 1816 and 1818 were years of grievous loss for lambs from snow and starvation; and in 1860 they were taken off the braes altogether to the low countries, and then they travelled with the greatest difficulty.

We had no such difficulty, as we fell in with a good lift back from the braes to Callander, and then spent the afternoon in a sort of mountainous journey past the "Old Woman's Burn" (so called from one who was drowned there), after M'Laren's cow which was first in her Highland Society class at Perth, and first also at Glasgow as a heifer. Mr. Gourlay Steell selected her as the most perfect female type of the breed for the gallery of the Highland Society, and he has done full justice to this buxom mountain mother with the yellow hair and the three white legs. Her daughter has taken prizes as well, but she has bred, so far, nothing but bulls, and between them they made up quite a red and yellow clan M'Laren, of which any man might be proud.

Deanstown is not far from the rail between Callander and Dunblane. Well nigh thirty years before, we had been there, when Mr. Smith was in the flush of success with his subsoil plough; but, boylike, we attended more to his luncheon than his lectures, and we remember little beyond than his bustling ways, his sturdy, little figure, and the glass frames in the flower-garden, which acted as skylights to the workshops below. This was in '35; and three-and-twenty years before that he had brought out his reaping machine. The first was too weighty, and the second broke part of its wheel-work by a sudden jolt in a hollow during the trial. Hence he did not win the £500 given by the Dalkeith Farmers' Club, but was consoled with a tenth of that sum for "meritorious endeavours." The Highland Society awarded him a plate for it, and so did the Emperor of Russia. Earth and water were alike his care. Draining and subsoil ploughs will always be associated with his name; he devised a salmon ladder with a 30-feet ascent; and then, with true cosmopolitan fervour, he invented and patented a dip for sheep.

Argaty, the old home of that fine prize-bull Van Tromp, is not far from here; and on Doune Green there stood the green sentry-like boxes of the bankers, awaiting the last tryst, and the skeletons of the tents wherein toddy has been drunk and the handsel given on many a flourishing day.

The carse of Stirling, where St. George's cross met St. Andrew's in

> "As red and rude a fray,
> As e'er was proof of soldier's thew,
> Or theme for minstrel's lay,"

well nigh six hundred autumns ago, was a land of mist that evening. We could scarcely discern the proud old Castle and Craig landmarks, which still bear their mute witness to the meeting of those mighty hosts, when the ploughshare no longer turns up the rusty pike-head, and even the memory of the "battle sheaves" has died away; but Dunmyot shrouded itself less sullenly behind us, and we caught occasional glimpses of the range of the Sheriff-muir Hills, with the little homestead of Pendrich clinging to their side. Gallows Hill, where many a "neck verse" has been sung, was on our right, and then a ride of a mile and a-half brought us to Keir Mains, the residence of Mr. Alexander Young, factor to the estate. Keir includes about 1,000 acres within its policy wall, and a large portion of it is in Perthshire (the Yorkshire of Scotland), which Mr. Stirling has represented since 1852.

The white nose of Peggy, was stretched affectionately over the hurdles in the beech-grove paddock to answer Mr. Young's greeting, as we started on our round. It was a lucky moment for him when he espied her as a two-year-old at her breeder's, Mr. White of Renfrewshire, as she has never been second save at Kelso, and completed her score of firsts at Stirling. A row of neat ploughmen's cottages, with a butt, ben, and middle room, face the entrance lodge to the steading, which is kept by the head shepherd. The weighing machine is in the hands of his deputy, and thus all the oilcake and artificial manures can be checked on their arrival. On the left is the stack yard, under whose wall the Scotch ploughs, all radiant with green paint, stand in file, along with rollers, grubbers, reaping machines, and haymakers. The steading is built round a court-yard; every division and passage in it has a separate letter, and A to Z exactly suffices. So complete are the arrangements that the very warmth of the engine-house is turned to account, and the hens are lodged over it, with a good plating of zinc between, to foil the "lively fleas;" while the engine is so close to the smithy that the blacksmith can manage it by means of a handle and an index.

The freestone of the country has been used, and the floors are laid with Arbroath flags, neatly and closely pointed upon six inches of stone metal. Nearly the whole of the interior walls are lined, like the tower, for about six feet, and beyond that with squared rubble, which avoids the necessity of plaster, and does not afford vermin the faintest hope of gaining a settlement. The principal entrance is at the north-east corner, near the granary. Beneath it, on the east range, are the cart-shed fitted with grease-pot niches, the barn, the straw-house; and behind them the sawpit, the smithy, and the engine-house, where a gigantic flywheel works an eight-horse power engine. At the corner of the south range is a twelve-stall stable, eleven feet high, and divided by a harness-room. Future Robin Burnses could have no more comfortable ingle to ruminate by, as the fire flashes cheerfully on the polished pine wainscoting; and the bump of order among the six must be feebly developed indeed, if with so many aids to neatness on the walls, they cannot keep their harness bright, and in its own domain.

Each of them has a separate bin, holding the exact weekly allowance of oats and beans. The stalls are nine feet by six, and iron-bound; the water (which is drawn from cocks in the stable), hay, and oats are all kept strictly to their three compartments in the manger, where the rope of the halter acts in a long square tube, and is untied from below. The overseer's house forms the centre of this east range, and is connected with the dairy, in which the delf dishes rest on large Arbroath flags, and the Guernsey milk is placed in tin bowls as being an essential cream-raiser. Polished tiles from the Hague, which represent a girl milking, a cock and hen, the Ranz des

Vaches, "Genesins," and Noah, are let into the walls, and the gallery above gives us a nearer view of fully 200 Moorish, Persian, and Italian plates, which are hung up like shields below the ribbon motto which encircles the ceiling : *" Cleanliness is next to godliness."*

Next in order are some loose-boxes, including one without corners for the especial use of a mare when she is about to foal. After a tour up the clock-tower stairs, at the south-west corner, to the future business room, we descended to the byre, which composes, with the cowman's house, the whole of the west range. The byre itself is 128 feet 6 inches long, and is so constructed that it may very easily be divided into three. It is built for thirty-two cows, which stand, two-and-two, in stalls of polished flag, each with an iron rack of its own, which is filled by the aid of the tram-road in front. On the north side are the piggeries, the calf-houses, and the lumber-shed; and fifteen loose boxes and two boiling and store-houses stand in the centre of the square, and encircle the manure-heap. The latter is so constructed as to be covered or not at pleasure; and the contents of the adjacent urine-tank can be pumped on it by means of the engine.

The size of the homestead may be judged of from the fact that the width of the roadway in the court, which will be paved with squared whinstone, is twenty-four feet. Of this, however, a small portion will be taken up with the tram-road, which will run round it, and thus open up a complete communication between the byre and the dung-heap. The pigs which revel there, when they are let out from the sties which face due south, are principally small white Wenlocks crossed with Wainman boars; and a wreckling was at nurse in "the bait" house, and trembling with anxiety and emotion when the teapot was presented. Among the foals in the boxes was a bay filly, the living image of Champion of the good old Shacabac, Fireaway, and Phenomenon blood. Vesta's last legacy, a red bull, was in the calf-house, where Forth's deep flesh was transmitted to a white son; and there was the first dividend, another roan heifer, out of the 235-guinea "Another Roan Duchess." The calves are all brought up by pail, and have two Scotch pints of new milk night and morning, and after the first month (when they run in the paddock if fine), boiled linseed is given them in their milk, and oilcake with bruised grain or bean meal at mid-day. The milk is taken off at the end of four months, and the linseed gruel continued.

There is nothing peculiarly decorative about the buildings, with the exception of the tower, but devices and mottoes have been applied with no niggard hand. Each department bears an emblem of its contents in stone; and hence a stranger wandering only round the outside can get all the bearings exactly. A horse's head—for which neither Clyde, Young Champion, Darkie, nor Audubon, son of Birdcatcher, could have stood to the sculptor—marks the stable; but Stella's and Lady Bountiful's heads have been copied by him for the cowhouse, and Forth's, cut in his calfhood, was over the keystone of the main gateway. A wheat and barley-sheaf crown the windows of the granary; and a brace of reaping-hooks and scythe-stones, crossed, are reserved for the door through which the sheaves are borne to the threshing machine. Mr. Stirling has well acted up to his family motto, "*Gang Forward,*" which once or twice had its place with "*Poco à Poco*" (Little by little) on the walls; and "*Tak Time ere Time be tent*" speaks with apt and homely eloquence from the clock-tower.

The stud of Clydesdales has numbered sixty, but it is kept down at about half-a-hundred, as the colt foals are sold entire, along with Leicesters and shorthorn bull-calves at the annual roup. Some of the neighbouring farmers thought Darkie, who was a good winner under high-weights at The Loo and Eglinton Park, too high-bred for their common mares, and therefore Mr. Stirling bought a pure-bred Suffolk sire; but the chesnut "bare legs" made no way. They said that the sort had too much roundness of bone, and a lack of freedom of step, and that the mists affected their eye-sight. The national feeling was also against them on the point of not being such good travellers, either as regards pace or the power of long-fasting; and the fifth Lord Jersey, who had tried large teams of both breeds at Middleton in Oxfordshire, was of the same opinion on these two points. The beginning of the stud dates from Clyde, who was purchased from Mr. Samuel Clark of Kilbarchan, and afterwards sold to the Speaker. He won a first prize when the Highland Society met at Glasgow in '50, and his beautiful head and three white legs were known in many a Scottish show-yard. His son Forth was third at Battersea and first at the Four Counties, and Baronet and eight mares and fillies have all taken first National prizes. Snip, Sally, Nancy, and Peggy were first-prize mares at Carlisle, Edinburgh, Battersea, and Stirling. Old Bet has never been in the Highland Society lists; but she has not shirked good company at Glasgow, Hamilton, and elsewhere, and has been decorated on sixteen occasions.

It is perhaps the prettiest farm sight in Scotland to walk down the Keir stable, when the horses are in from work and all done up on an autumn evening, and a robin redbreast twittering on one of the stalls lent it extra zest. Douglas's Snip, the winner of three Nationals in her day, stood at the end; but although the fine fore-arm and leg are still there, a decade of years have told their tale, since Ralph led her out to victory at Carlisle. She was the only one which had crossed over the Irish Channel in the legitimate pursuit of Ribbonism; but every one of the twelve, save a big half-bred black of the dray order, was a winner. Punch, with that astounding rib, is old Snip's partner. He was bought by Mr. Young at Banff, when he was judging with Professor Dick, whose lament over his not being in the sire list was at once pathetic and incisive. Star is still more to our mind, with his grand quarters and jet-black legs, and he won in a great ring at Ayr. His Platonic mate is the buxom Lily, beautifully turned, but still not quite so orthodox in her shapes, and with a rich ruby-coloured coat that rhymes but ill with the mealy bay of Jess, which has, however, not kept her back from honours. Sally and Bet were in the boxes, and scarcely do any work now; but Mally, a bright bay with white hind-legs, is in her prime, and with few to match her for style. She pursues her way with Katey since Bessy was sold into an Edinburgh lorry; Duchess and Polly are closely akin to Blackleg, once quite a terror to the show-yards; and a little pair, Nance and Bell, cast in their lot together. They have always a pail of "bait"—or rather turnips (green-tops in autumn, and swedes in spring), chaff, and a little barley, all boiled together—when they come in about four, and at eight each night the watchman gives them a few raw turnips, to amuse themselves with and gather sleekness withal.

There was good stock at Keir in the time of James Stirling, the late laird and uncle to the present. He had a catholic feeling for every kind of animal, but more especially for shorthorns and greyhounds; and no one fed West Highlanders to higher Christmas weights. He brought "the Durhams" into Perthshire, and stood by them manfully, when the Highlanders shook their heads sagely over "those big, painted beasts;" and every day during the season he might be seen on the carse, with a brace of "long-tails" and an Argus-eyed groom at his heels. The stock was sold off in '45, and, seven years after, his nephew came out with fat beasts, and disposed of the three great Aberdeen butchers, Martin, Stewart, and Knowles, with a Shorthorn-West Highland heifer. After this, the Keir showing took a breeding stock turn, and Shorthorns, Leicesters, pigs, and Clydesdales have gathered nearly one hundred firsts between them, principally at the Royal English, the Highland Society, Glasgow, the Four Counties, and Stirling.

The present herd, of which Blencow (11182) of the Gwynne tribe and four or five of the Sweetlips from Boswell's of Balmuto were the germ, dates from '52. Leader (11674), bred by Milne of Faldonside, was the successor of Blencow, and then Fawkes's John o'Groat (13090) by Bridegroom (11203) joined the herd for 200 gs., after he had been placed second to Master Butterfly at Carlisle. The two roans did not meet at Malton, where "John" was sent by virtue of a stipulation, and the first honours which he won there were confirmed by the silver medal at the Glasgow Society for the best bull in the yard. Next summer he stood first at Salisbury, and earned from Mr. Wetherell, who is by no means diffuse in his panegyrics, the title of "*the finest big bull I have ever seen.*" We cared very little about cattle in those days; but we happened to be at his stall when "Nestor" came up to handle and deliver judgment on the roan. After winning at the Highland Society that summer, he was seen in public no more, and was swept down by pleura along with two-and-twenty of his mates. He left a few heifers behind him, and in the roup of the Salisbury year his Marble Cutter from a Blencow heifer made 200 gs., and went to Australia shortly afterwards at an advance. Sir Samuel would have come for a season from Warlaby, but the pleura was too recent to risk the "last slice of Charity;" and Hiawatha (14705), by Captain Balco (12546) from Playful, who had been purchased from Mr. Douglas the year before, took the vacant box. Previously to his purchase, he had beaten Colonel Towneley's Fred and Sparrow Hawk, Mr. Ambler's Museum, Sir James the Rose, and a rare row of bull-calves at "The Yorkshire," and in Mr. Stirling's hands he beat the class of young bulls at Glasgow, when John o'Groat headed the seniors, and Miss Nightingale was third to Rose of Athelstane and Ringlet. The Keir bull-box has never lacked a great prize-winner, and it is remarkable that on the only three occasions that Mr. Stirling's bulls, to wit, John o'Groat, Forth, and Eleventh Royal Butterfly, have been across the Border, they won a Royal first.

The original steading has been gradually absorbed into the pleasure-grounds, and the site is occupied by a small cupola, and planted with Turkey oaks and variegated planes. The beautiful dish-head and the Godolphin crest of Champion are laid low, to the deep grief of Tom Liddle, after ten seasons; and his old box in "Clydesdale Lodge" (which has a large high-walled yard attached to it for exercise) is filled by a Clydesdale of the lighter sort, and fully a hand less than Baronet by Rob Roy, who came from Renfrewshire. In the home-field below, Duncan in his Scotch bonnet was leading the lengthy, fine-loined and short-legged Eleventh Royal Butterfly for his morning constitutional; and he might well be proud of the office, for Towneley has bred but two to compare with the high-mettled roan. The little man totally despises poles, and manages his 400-guinea charge in all his caracoles with nothing but a rope and a small cart-whip. The latter is in strict keeping, as he always addresses him, either for warning or encouragement, in strictly cart-horse language. The capricious Master Groat and Knight of the Border were not his peculiar charge; John O'Groat and his 150 Dutch stone he could mould like wax; on the eve of harvest festivals he has often taken a nap in Forth's box, and the bull only licked him as he slumbered; but "The Royal" is not to be trusted so implicitly, and therefore Duncan guides his slumbers with discretion.

There are only two John O'Groat's among the herd which is grazing beneath us, and both of them good to know from that peculiar mode of poking out their heads, which they derived from the old bull.

Miss Groat is one, and so is Anna Rose, the dam of Forth, and so ludicrously like him in shape and colour that we found her as easy to guess as we did the dam of Sir Richard at poor Tom Rea's. Annie, Wellingtonia, and Lady Airlie form a deep red trio with all the traces of their sire Hiawatha in the head and the set of the horn. Minnehaha by Heir-at-law was there with her broken horn, and so was Miss Wetherell, the only relic of Windsor Flower, and the only calf, save Forth, that Florist left behind him. The sweet, short-legged Vesta, who kept such high company in the days of Frederick's Fidelity, Rosette, and Queen Mab has gone the way of Windsor Flower; but the slashing Miss Nightingale by Grand Turk, who was also just over-weighted in the prize-lists, was there, "with milk for any two"; and a daughter, Nursery Maid, to speak for her as well.

Mr. Binning Home's Van Tromp, Forth's strongest opponent, has his say in White Rose; Baroness Cherry represents the Roan Cherry tribe; and Heiress of Killerby the "single speech" prowess of the Heir of that ilk. Rosy from Syme of Redkirk had just headed the last roup with Knight of Stirling (75 gs.), and Princess of Cambridge the preceding one of '63 with her Allan (92 gs.). Winning Witch, like Vesta, recalls "Tallant and Bushey"; Another Roan Duchess with the unmistakeable Frederick roan carries us back in the spirit to the great days of Towneley and her invincible dam; Mysie 13th keeps up

Y

the honours of a useful tribe; Princess of Cambridge strains on her sire's side to Bolden's Grand Duke; and Maid of Athelstane wanders a maid forlorn.

A few of the Leicester ewe flock are in the Home field as well, and among them several black ones, selected by Mr. Young out of the forty which were sold by the late Mr. Boswell at his last roup, with his other fancies, the Shetland pony pairs, Semibreve and Octave, Tivy and Tantivy, Gippy and Tippy. About 40 Leicester tup lambs are brought into the roup ring on the slope near the main gates each October, and mostly go north of the Forth to cross the black-faced ewes. The last average was £4 3s. all round, and several of them have made £5, £7, and £10. The Leicester blood is a combination of Lord Polwarth's with Cockburn's, Simson's of Blainslie, and Bosanquet's; and there is also a slight infusion from Brown of Burton, Roy of Nenthorn, and Carter of Richmond.

Now we peep into the heather-covered platforms, where a century of peacocks roost at night, and take a round among the bullocks, Leicesters, and Clydesdale fillies in the park. The north wind had stripped the leaf from the beech, and was whistling through the tassels of the larch; and the silver ball reflects nothing but wintry barrenness in that carse, in which Dean Stanley saw such a vivid likeness to the Plain of Sanur, but Keir is still green and fair. You can still ramble among thick laurel hedges with standards, silver and golden hollies, costly deodaras, and cedars of Lebanon; and *"Homo quasi flos egreditur et conteritur"* whispers its warning in sea-pinks as you wend your way to the Rhymer's Glen.

CHAPTER XV.
KEIR TO FIFE KENNELS.

"A filthy beast, Sir! Why, a cow is one of the most agreeable of all animals. Everything about her is wholesome and useful: we get odour from her breath: she supplies our table with meat and butter and cream and cheese; and I assure you, Sir, I would rather eat a cow than a Christian."

ROWLAND HILL.

Alloa—Mr. Mitchell's Herd—To Keavil—The Keavil Herd—A word with Mr. Easton—The Old Fife Breed of Cows—Fifeshire Feeders—Old Days of the Fife Hunt—A Visit to the Kennels.

THE road from Stirling to Alloa lies through the strong wheat lands at the foot of the Ochil hills, which then bend sharply away to the left towards Perthshire. Some of the sheep-ranges are 2,300 feet above the sea-level, and were once held solely by the black-faces, but the Cheviot usurper is fast gaining on them. The grass is peculiarly healthy for sheep, but the farmers do not care to feed, and therefore sell off their Cheviot wedder lambs. A few keep blackfaces to stock their "wedder ground," but all such distinctions are fast fading before the high price which lambs command.

Alloa is deeply devoted to ale, whisky, and wool. The water rises from a freestone stratum, and has strong functions to perform. Nearly three million gallons of whisky are made annually in Clackmannanshire, and so are ninety thousand barrels of ale, which go direct to Liverpool, London, Glasgow, and Newcastle. About 150,000 quarters of barley and other grain, principally from the Mediterranean and Black Sea, are used for whisky alone, and 22,000 of barley for ale. Australian, German, and Scottish wools are all used in the manufactories, and are chiefly worked into tartans, stocking-yarn, and tweeds.

Very few pure sheep are bred in the low parts of the county; but the Leicester tup is crossed with Cheviots or black-faced, as the pasture suits; and the lambs are sold to Fifeshire or the Lothians, either off the farms or at the Glendevon market at the east end of the Ochils. Many of the farmers buy half-bred lambs at St. Boswells, and sell them, generally out of the wool, between April and July; and those who get cast Cheviot ewes, keep a small flock of Leicesters, and breed enough tups for their own use and to exchange with their neighbours. The land, which grows better wheat than barley, requires much labour to prepare for the turnip crop, and is very trying to the Clydesdales. All along the Carse of Clackmannan it is a great bean country, and the lasses in their white crazies, one drawing and the other handling the drill, make a brave show as they follow the plough in March. Scarcely a hunter or roadster is bred in the county, although the farmers have had the choice both of Physician or Liverpool blood, with Æsculapius and Moss Trooper. There are a few West Highlanders and Ayrshires, and the "Falkirks" are handy for yearling Irish stirks. Cross-bred country cows are put to shorthorn bulls; but no one in the county, with the exception of the Messrs. Mitchell, keeps a shorthorn herd.

The brothers began about the time of the Chester Royal, and have had Prince Arthur, Sir Colin, First Fruits, and Sir Samuel on hire in succession from Warlaby; while the herds of Messrs. Gulland, Troutbeck, Crawley, Steward, Jolly, Wood, Milne, Spencer, and Towneley have furnished most of the females. Except at the Highland Society, the United Counties of Perth, Fife, Kinross, and Clackmannan, Stirling and Dunbarton, they show very seldom, but invariably with success. Their farms, which lie partly round Alloa, and partly on the Carse of Clackmannan and the higher land, comprise 1,200 acres; and what with his malting, farming, shipping, milling, and coal mine evidence, we remember Andrew, the elder of the two, puzzling a House of Commons committee not a little as to the exact nature of his profession.

Mistletoe is his herd matron. She cost 74 gs. at the Crawley sale, and as a yearling and two-year-old she won seven prizes. One of them was gained by lapse at Perth, when Soldier's Bride was disqualified, and when the "unco' wise" prophesied in print of Mistletoe that she would never breed. At York she was third to Queen of the Ocean, and Pride of Southwicke, the first-prize Royal cows of 1862-63, but she slipped her calf on reaching home. Her Conqueror by Sir Colin was first in the yearling bull class the next summer at Kelso; and then the old cow not only beat a very strong class at Stirling, but claimed the prize on the birth of her fourth calf. So much for the seers!

We watched part of the herd—which is some thirty strong, without the calves, and all but three or four by Booth bulls—filing in from Mars Hill past the new Alloa Hall of Justice to the steading which lies on one side of the town. The ten-year-old Sir Samuel, whom Richard Booth loved so dearly, not only for his fine handling, but for Charity's sake, that he never let him but once, stood in the first box, next to Lady Laura, who has something of Queen of the Vale about her, and always played second as a two-year-old to Mistletoe. White Eagle, a lengthy cow by Knight of Warlaby, and whose dam, Lady Eagle, was bought for 105 gs. at the late Captain Spencer's sale, the nice-haired Comely 3rd by First Fruits from one of Mr. Nicol Milne's tribes, Luna, another First Fruits with a very neat leg and going back like the big red Lady of the Lake to Jobson's sort, were in one byre, along with Barbelle, a neatish cow with a curious horn, and a 74-guinea prima-donna at Mr. Wood's sale.

The substantial Nervosa Booth, by Prince Arthur, and her half-sister Cameron Lass with her fourth calf at five years old stood in another which had four Prince Arthurs in its six stalls. There have been three sets of twins (of which five lived) in the herd within twelve months, and two sets came out of this byre. Mistletoe, a very deep-fleshed, robust-looking cow by Welcome Guest from a Grand Turk dam, has not followed this example, and unfortunately she always breeds bulls. Her Stonehenge, who has since been sold to Sir George Dunbar, was on parade in front of her box along with Thane of Fife, both of them with first honours from the United Counties; and her youngest hope, Red Friar, and Lord Eagle, out of White Eagle, were surveying them. Facing the white Cherry by M'Turk, a compact, good-looking cow, and side-by-side with the fine-boned Pauline by Highthorn, was the neat, wide-spread form of Blue Belle, safe in calf to Sir Samuel. She and Eagle's Plume were the first that the Messrs. Mitchell ever sent to the Royal, and they returned with a second and a commended ticket, and drew up into the first and second places at Stirling. Breeders have had their doubts as to which is the best, but generally agree that, if the white has more length, she has not quite the width and sweetness of the roan. Let us trust that it will be long before they share the fate of old Nervosa Gynne, whose Arthur Gynne was used both here and at Keir, as she was busy in the darkest and farthest corner of the yard, laying on beef for the flesher.

Clackmannanshire is a very Rutland among counties, as you see pretty nearly the whole of it in a ten-mile ride along the Carse and the Forth side to Keavil. The capital is not impressive. There is a clock with a cock on the top of it, and a Druidical stone with a cross, at the base of which Robert Bruce is supposed to have tied his shoes. A steamer is tugging up a brig with a rich cargo of grain and groceries past the mouth of the old coal-mine, whose shaft looks like a ruined abbey overtaken by the tide. Bretonnes and Alderneys are in possession of the grass Park of Comte de Flahault's and Lady Keith's at Tulliallan, and the rich traces of their presence are to be found in its cheese. The West Highlander is also no unfrequent tenant of the pastures, as we ride on past the gates of Torriburn, which furnished both the best blood sire (though one tinge of the dreaded *h. b.* made the decision void) and the head of the polled-cow class to the Highland Society at Kelso. Pretty beech glens fringe the road to Keavil, that home of Englishman and Seraphine 13th, and at last we see Mr. Barclay's snug grange half-hid in the oaks and planes "standing there in ages gone by," as Mr. Easton, the bailiff, observes. Our old friend was looking round in his hat of Leghorn straw, and was as full as ever of those dry aphorisms which have so often tickled the show-yard and the ring-side, where he is always so marked in his attentions to primadonnas. Two weeks earlier, and we might have seen him looking like a perfect Boaz among the oat sheaves which had given such fruits of increase.

But the summer was past, and he was walking among a troop of Leicester ewes in the paddock, like a huntsman with his hounds. "*Come awa', come awa', my wee doddies!"* he says; and even if they had not been hand-fed, they could not resist such blandishments. *" Thirty ewes and their production"* compose the Leicester flock, which is principally a cross between Lord Polwarth's and Mr. Cockburn's of Sisterpath. The "production" had been 51 from 26 that season; and each year the tups go to the Edinburgh tup sale, or are sold at home. In the best year so far, twenty-two of them without the fleece averaged £5 15s. 4d. Thanks to the shelter and high feeding, Mr. Barclay calculates the fleece of the ewe hoggs at 10lbs., the ewes at 8lbs., and the tup hoggs at 11lbs. The very sheds are worked on a regular rotation, and when the calves have done with them, the hoggs take their place, and eat turnips under shelter in the frosty nights. The same sort of scene-shifting, but of a more elaborate character, takes place in the stables, which are large enough for a master of hounds; and with merely a manger alteration and a chain-pole, they help out the byres in the winter. There are about forty old and young in the herd; and Captain Gunter's Northern Duke by Duke of Wetherby, and Mr. Bruere's Baron Booth by Prince George, have been purchased as the representative bulls of Bates and Booth.

Faith occupied a capital loose box made up by the union of two stalls, and the long and low Prudence looked as beautiful (though of course small in her horse-stall setting) as Little Lady, when that artistic light, in which Lord Stamford delighted, was wont to fall upon the bay in her Newmarket stable. The hens and turkeys are a great point here, and seem to have a most lordly time of it among the Portugal laurels. Turkey cocks and hens, by a sort of mysterious etiquette, separate towards the middle of the day, and sit demurely on separate rails. Mr. Barclay is as choice in these matters as he is in every thing else, both inside and outside his house, and adheres rigidly to the bronze-grey sort of Cambridgeshire. The Norfolk cross rather spoilt his size, and connoisseurs do say that the small black turkey from the North of France has done that county breed no good. At Keavil they average, on their Indian corn, meal, rice, and potatoes, about 19lbs. all round at 6½ months, but many of them are gradually killed off as poults.

But Mr. Easton will have us away among his "sappy queens," and we are not loath to obey the call. "There," he observes, "is Frontlet from Mr. Adkins; Miss Burdett Coutts, the one nearest you. The twins, red and roan, you remember, only the roan has lost a piece of her tail. They both had calves when they were only twenty-five months old. They're always together. The roan's milked three times a day: we're a little kind to her." .... Yes, Mr. Culshaw always says that he thinks '*he'll give them a little something to eat.*' Well, well! I dare say we both mean the same. Mr. Culshaw's a man of discernment. The roan's Lady Mary and the red 's Lady Anne: they are from Cruickshank's

Lancaster 25th by Lord Raglan. She had a triplet; it's a fact not more curious than true. The third, a roan, came seven hours after, when we had bedded the cow up for the night, and left her; it was the finest of them all, but it got smothered. At Perth the red beat the roan, but the Kelso gentlemen reversed it. '*Come awa,' my Seraphine 13th!*' I met Captain Oliver lately, and he asked very kindly after her. He was our opponent at Southcote. He's not easily beaten off. I said to Mr. Barclay: '*Go in, and give the Captain another choker: it's as well to do right as wrong.*' So we got her for 250 gs. She will be just three years and nine months, and she's had two calves, and in-calf again. She was beaten at Stirling, and people came up to me, and asked if it was right; but on those occasions I sit down in my corner and say nothing: it's the best way: talking doesn't avail. *She was seen.* We never fed her for it; she was only led about this field for canniness.

"There's Water Maid, a nice deep roan of the old stamp, a great favourite of ours. She was the *prima donna* of the Maynard sale; Mr. Barclay gave 110 gs. for her. We have her portrait, as true a touch as I ever saw. What 'a plateau,' as they call it, she has over the loins! We've given Englishman the best of the cows. We like to pay respect to the animal, and the man who had him. Those are both English ladies—one from Sylph: '*Come to me, my darling! Sylphida—that's your name. Come, my gentle queen!*' This is Flower of Spring. No, no! we've not forgotten Emperor of Hindostan; we've six by him. There's Lancaster 25th, with her son in the home park, under the plane-trees, called sycamores in the days of old. That's a neat little roan by Englishman out of Prudence. There's Platina, a white: we've just enough whites; and we've got Royal Errant, who beat the Royal Newcastle bull, for her, and to help the white into red. The Duke was very gracious about it. Royal Errant's bred like Blue Belle—from a Cardigan cow.

"We'll get over into the next field. You jumped those iron rails well for a family man, and yet you must be tolerably stiff with all these peregrinations. Do you ever sleep at all? That's Lady Raglan, dam of Flower of Spring. We'll just take the outside ones, and work to a centre. Here's Sylph, one of Dudding's sort— one of the Sir Samuels, and second at Lincoln Show. That's your old friend Prudence by High Sheriff: you liked her in the stable last year. They've fat backs, but they've only the pullings of the field and good constitutions to digest them. This is Annie Laurie. '*Come awa'*, *Annie!*' She's a gay young lady, as wild as a Highlander. Englishman and her will make a fine cross. Faith's always by herself; she would still be lonely at Mr. Sanday's. She broke down on us with a premature bull-calf. We hope better things of her: but she's all waves and hillocks to look at. Mr. Houseman can dissect her down to a sixteenth part. He dissected her very prettily a fortnight after we got her. He told us all

about her. Nonpareil by The Baron—that's her
Her breeder parted with her in an evil hour. Only
she and Lancaster 25th left out of that lot. The other
two, Duchess 2nd and Matchless, were barren, and
sold. If we had saved the triplet it would have
squared matters. Last, though not least, there's
Fan Fan, the white daughter of Sir James; she has
a calf, the image of Emperor, six months old.
That's the town of Dunfermline on the hill; the
abbey and the spires, and the damask manufactory;
three acres for one of them. Upon my word, they
are spreading their wings!

"This is the yard; it's quite a menagerie here.
There are no pig sales in these parts. Mr. Findlay
gets all the pig money. They think it, about here,
stiff enough at 20s. for one of eight weeks. There's
one sow of our own breeding, and the others are
Lord Wenlock's breed. That's the family! One is a
little too close bred, from a Wenlock and by a Wenlock; its tail came, and it withered.

"The bulls are this way. That's Silver Duke:
he's a nice waxy lad; but this is his conqueror!
We've mostly Bates, with a mixture of Booth for
emergencies. Here's a gentle thing, Emperor of
Hindostan, but he's got the lion's share of the cows.
He stands well in the book. We've a twelve days'
calf by him out of May Queen. Vine Dresser—look
how he stands round to let you see his chest. That's
an Englishman calf, and his dam's a beauty. We
think as much of his father as any body's bull: I'm

happy to say he's rallying: he puts them all right on the top. '*Englishman! here's an old friend come to see you.*' He always roars this way when any one comes in that he doesn't know. He was blistered on the chest and gullet. We have given him aconite to act on his heart, and now he gets *digitalis* and green food."

Not a Fife cow was to be found even as wet-nurse in the herd; and now that they have been struck out of the Highland Society's list, we did not care to search for them beyond the Society's picture. Some of them had brocky faces, and the popular belief is that they owed their origin to Germany. They are middle-sized horned blacks, not unlike the old Hamburgh breed, or it might be said a cross between an Ayrshire and an Angus, and alike good for the shambles and the dairy. Mr. Aiken of Carnbee had some of the last winners, and Mr. Stocks of Beveridge has still a dairy of them near Kinghorn. As a feeding county Fife stands very high, and pours out its beef supplies from February to June, with boundless plenty and precision. The Angus and Galloway beasts were once termed its "spring keepers," but the Shorthorns have gradually crept in during the last twenty years, and their quick feeding qualities have carried all before them. Its farmers breed very few sheep and cattle. They go to Melrose or St. Boswells for half and three-parts bred, as well as a few Cheviot lambs to winter; but a good many cross-breds are bought at Glendevon, off the Ochil Hills. Their own great markets are at Cupar and Kinross, and their best feeding beasts are bred in Forfarshire and Kinross-shire; but the bulk are bought at the Falkirks, and Hallow Fair. The largest feeders in Fife are the Duncans, Alexander, Thomas, and Robert of Pusk, Boghall, and Kirkmay, the Ballingalls of Ayton, and Dunbog, and Alexander Reid of Cruivie.

The *Sederunt,* which established the Fife foxhounds, was held on May 7th, 1805, at Cupar. Mr. Gillespie of Mountwhanny, Mr. Johnston, jun., of Lathrisk, Mr. Patullo of Balhouffie, and Mr. Dalzel of Lingo composed it. It was proposed to raise £800 a year for ten years; and £700 was promised in the room. General Wemyss, Sir W. Erskine, and Mr. J. A. Thomson gave £100 each, the Fife Hunt and four other members £50, and six other members, including the Sederunt, £25. The thin attendance was the cause of an adjournment for a month to the parlour of Mr. M'Claren, vintner, when Mr. Johnston was absent and Colonel Thomson present. There were some refusals, doubts, and "conversations" reported, but £75 more came in. Being armed with full powers, the second Sederunt went promptly to work. The Harrier Kennels at Brock Hill were inspected and approved of, a stable was built 30 feet by 14, and fresh land feued for a hound-yard from the Council of Cupar. A huntsman was advertised for in the York and Edinburgh newspapers, the whipper-in to the harriers was kept on, and David

Law was fixed on for feeder, with this proviso, "if he can be got." Three horses were purchased for £148 1s., and one was received as a compliment from Major Thomson, who was one of the first committee, with Sir Charles Halkett, Mr. R. Ferguson, Mr. Gillespie, Mr. Patullo, and Mr. Dalzel. The financial part of the question was grappled with in a most business-like style, as the Secretary was empowered, on the 2nd of November, 1805, to "charge interest upon such subscriptions as remain in arrear from this date." The hounds that were purchased only cost £54 8s., and the drafts were advertised in the Edinburgh papers. All these precautions did not avail, as on March 24th, 1809, there was a debt due to the treasurer of £563 7s., and only a very problematical chance of getting in two years' arrears of £100 to set against it. The next Sederunt compared their accounts with those of the East Lothian, and came to the determination to have "a total change of servants." The change was not for the better, and the debt swelled to nearly a couple of hundred more, and affairs became so puzzling that Mr. Rigg offered to keep the hounds for £800 a year. His offer was accepted, and the debt was more than half paid off when the first ten years had expired.

They were then established for another five years, "no subscription to be received under £25." General Wemyss, Mr. Christie of Durie, Mr. Rigg, Captain Hay, and Mr. Moncrieff became the committee of management, and £50 was " supposed to be voted by the Fife Hunt as formerly." It was also settled that Cupar and Dunfermline should be the stations for the twenty couple. The hounds were to go to the latter town after the October meeting of the Fife Hunt for as long as the gentlemen in that quarter wished to have them, to Cupar till after the spring meeting of the Hunt, and then finish up the season in Forfar or any neighbouring county the committee might appoint. So far, so good; but some hitch would seem to have arisen in the Dunfermline country, as, by a New Inn minute, the proprietors of coverts in Clackmannan and Kincardineshire were begged not to destroy foxes. Shortly after this a five years' arrangement was come to with the Perthshire men, through Sir David Moncrieff, to subscribe £480, and have the two Cupars and Bridge of Earn as the principal stations for the ensuing year. Towards the close of the season of 1820-21, Captain Douglas wrote to say that Lord Kintore was leaving the Forfar country, and that there was a wish in the county to combine the hunts. The committee were accordingly authorized to "confer with the gentlemen from Angus," and, as Mr. Rigg stated that "the game is very scarce in Fife," the hounds hunted in Forfar for the remainder of the season.

In 1821, the two packs were united, and two gentlemen from each county formed the Coalition Cabinet. The leading conditions ran thus: "No covers to be drawn North of Lawrencekirk or West of Belmont; and in case of a separation, the Fife men to draw twenty-five couple of the running hounds by ballot, and the Forfar fifteen, and divide the puppies." This arrangement went on for three years, at the end of which time the Forfar gentlemen were £795 15s. 5d. in arrear, and as it could not be recovered, £800 had to be borrowed, and the hounds were handed over by a deed of transfer to the gentlemen who had become answerable for it. The West Lothian men did much better, and being inclined for five weeks' sport in 1842, they offered £200 and paid it.

Having thus acquired a sort of national debt, things began of course to look up. On July 27th, 1827, Captain Wemyss and Mr. Whyte Melville were appointed joint managers, and after eleven seasons, the whole management was vested in the latter, by whose Gladstonian manipulation of ways and means and Will Crane's and John Walker's fine science as huntsmen the hounds were carried bravely on until the end of the 1847-48 season, when they were sold for £500 to Sir Richard Sutton, and Walker went to succeed Will Grice, or rather Jack Woodcock (who hunted the hounds for one season), at Wynnstay. Thus the old Fife Hunt ended with a balance of £40 9s. 5d., which was paid over at the beginning of 1850 to Earl Rosslyn on behalf of the New Fife fox-hounds.

A friend has sent us a slight *résumé* of the "Merry John" days: "Will Crane died in the middle of the 1829-30 season, and John Walker from Lord

Kintore's succeeded him, and was huntsman for eighteen seasons. Captain Wemyss found the horses and Mr. Whyte Melville the hounds. The kennels were at Cupar, and also at Torriburn, twenty-one miles from it, in the west of the country, where the hounds went for three weeks in the autumn, and three weeks in the spring. Dunfermline was then the head-quarters for the scarlets. In Fife there is a finer scent over the fallows than the grass. The east part is old grass. We could race in the west when the dust was flying, and we could do nothing in the centre or the east. There were great meets in the west, and the foxes never turned their heads. Mr. Ramsay's hunt used to join in then.

"The whole country is a fine mixture of plough, grass, and sheep-walks. We had some beautiful runs over heather, straight across the Dollar Hills. We had also some rare runs with Walker from Belliston, Stravithey Gorse, Kidd's Whin, Largo Law, Mount Melville, and Bishop's Gorse. One of our very best was from Scots-Craig, one hour and ten minutes, to Crail. Walker was close on his fox, and they ran him down to the water edge. We saw 'Charley's' ears twinkle, and then he swam out to sea 150 yards, with the hounds after him, and sank like a stone. The whole body of the pack swam round and round the place where he disappeared, and then gave it up; but Vaulter stopped and retrieved him. There was always a very large field on the Edinburgh or Kirkcaldy side, and the officers from Jock's Lodge barracks. Sir Hope Grant, Captain Percy Williams, and the 9th Lancers were there. John Walker used to say that there were sometimes nearly thirty in the field who would all make huntsmen or first-whips, and no small credit to him for the teaching he gave them.

"Lord Elcho and Sir David Baird were often with them, and so were Major Douglas and Lord Kintore; three Captains, Hay on Wasp, Wemyss on his bay Driver, and Wedderburn on his thoroughbreds, all went well; and so did Mr. Gillespie of Mount Quhanny, Lord Rothes of Leslie (a light weight), the two Stewarts of St. Fort, Mr. Balfour, Mr. Whyte Melville on Malvern, General Lawrenson, and others still going or gone. Tom Smith, Jack Jones, Stephen Goodall, and Cooper, all whipped in to Walker, whose favourite horses were Grocer, Doctor, Kitty, Lucy, Clinker, Farmer, Major, and the grey mare Nutmeg. There were no hounds for one season after the pack were sold; but Mr. John Grant of Kilgraston hunted Perthshire occasionally from Sir David Moncrieff's kennels, near Bridge of Earn, and then sold his hounds to Sir Watkin. Then Captain Thomson kept hounds for one season at Charleton. He had the Donnington dog pack, which he purchased from Mr. John Story and Mr. Seymour Blane. Will Skene, one of Walker's disciples, was head man, and Charles Pike from the Devon whipped in. They were very short of foxes, and the hounds run roe-deer like fury. Many of them went back with the Captain to Atherstone, and they turned out very well.

"Lord Rosslyn established the present pack, and managed it with Mr. Balfour of Balbirnie, Mr. Oswald, Mr. Peter Paterson, Mr. John Haig, and Mr. Frederick Wedderburn, for six seasons; while Captain Thomson was with the Atherstone, and in 1858 Captain Thomson took to them again, along with Lord Rosslyn, for a season, and had his first day in Melville Woods. Lord Rosslyn joined Lord Derby's administration, and went to the War Office the next season; and Captain Thomson carried them on up to the spring of '64. He then went to the Pytcheley, and under the next master, Col. Babington, the country was hunted five days a fortnight, instead of three days a week. At the close of last season Mr. Balfour of Balbirnie took them for two seasons, so that we have had many changes of Ministry."

The New Inn at Fruchie Fife has been for sixty years a favourite tryst for the Sederunt, and it rose in 1850 to kennel dignity. Potts, Jack Grant, and Oxtoby were all huntsmen there in turn, and Captain Thomson put up Fred Turpin, who had whipped in, and then hunted the hounds, in consequence of Oxtoby's illness, the last season. Fred completed his fifth season with the Captain, and then they parted, one to the Pytcheley and the other to the Vale of White Horse. The Inn is only one in name now, and "the iron horse" runs close past it, and mocks the fate it created. It lies pretty nearly in the centre of the country, and it is only necessary to go to Nottingham, two fields away, in order to find a fox. Lomond Hill is hardly two miles off, and when it does require routing, about four times a season, every hound that is fit bears part. Forty-four couple were in it one day, and after working among six brace of foxes for eight hours, a brace were killed, and a brace marked to ground.

Its hunting stable was principally filled with Irish horses bought in Perth. "The Dentist" had knocked out the teeth of a dealer, for a standing testimony against ginger. There, too, were Snapdragon, Crinoline (a fine lengthy mare, but not so dear to Fred as Kathleen), the big Victor Emmanuel, the lop-eared chesnut Kangaroo (and a trimmer if he were not touched in the wind), and Ben with a knee bound up, which "is no disgrace to a Fife hunter." Strange horses always cut themselves on the curb place in Fife. The ditch on the taking-off side is $1\frac{1}{2}$ yards from the wall, and if they drop short they don't get their hocks over. Walker always obviated this by going as hard as he could, and clearing everything. Captain Thomson's horses were at Charleton, and from the Lothians they can be seen with a good telescope, at exercise along the sands. The slashing chesnut "Highlander," bought out of a drove at Brechin, would be good to tell at that distance, if sixteen-three and six feet and five have anything to do with it; and so would Gladiator and the six-season grey Unicorn. Highlander was by Ferneley, the property of Lord

Strathmore, out of an Arab half-bred mare which ran in the Defiance coach. Phœbus, the sorrel stallion by a Norfolk Phenomenon horse out of a thoroughbred mare, was as stout as a castle, and as clever as a cat, and so was a horse pony from the Atherstone country, where his owner had been laid up for two months with a broken bone in his leg.

The heads of Syren and of Benefit, that mother of the Gracchi, adorned Fred's chimney-piece, with photographs of his father-in-law Will Danby, and Blossom, while the skin of the big Blucher hung like a mantle over his chair. Benefit, by Burton Comus from their Benefit, came originally in a draft from Dick Burton. She was so bad in the distemper that Mr. Henley Greaves thought she was not worth carrying away, but Captain Thomson took her in the Burton draft along with the Donnington dog pack of 32 couple, to Fife in '49. Her Blossom by Atherstone Ravisher was an especial pet, and the Captain only retained her and her sort, when he sold off at Stratton Audley in '57. She had fourteen puppies that season to Morrell's Bajazet, and four couple of them were still running to head in their fourth season, and among them Bonny Lass, "the largest and most powerful combined with quality" that her owner ever bred. Her nose was the same to the last even in her tenth season, and "her lovely eyes" were as bright as ever. Blucher was the solitary puppy of her old age, and she died in whelping him. In one of her litters she had seventeen, and with wet-nurse aid she reared sixteen. Fertility runs in the sort, as she had forty-one in three successive litters, and Trifle her daughter died with sixteen. Syren was not so shy in the field as she was at the Guisboro' Show. During the season after her victory there, she came cantering to meet Captain Thomson as he rode up to the meet at Mount Melville, and happening to touch his horse's hock with her nose, the white lashed out and killed her on the spot.

Once upon a time, hounds could always make a fight over Fife, but now drains and guano tell their tale only too surely. It is a good scenting country, but four-fifths of it is on the plough. There is no hedge-row timber, so that you see mischief before you, but plenty of brooks and ravines, which require a man to know the handy places, or disappear then and there from the front. From east to west—Cambo to Shaw Park— it can be very little short of sixty miles, and there is a vast expanse between the Tay and Forth from Queen's Ferry to Kilgraston. The east, of which the best part is from Ceres eastward, is mostly old pasture, and its farmers ride far harder than the west country men. The foxes often run along the sands, and then sit down under ledges of the rock, and the hounds work it out, feeling like their rough-haired brethren, for the scent in the waves, and speaking to it here and there on a stone. Most of the covers are young plantations with gorse, but they are getting very hollow at the bottom. Kidd's Whin belongs to Captain Thomson, and was made by Walker during its owner's schooldays in four bits of ten or twelve acres each; and it was there that "Merry John" claims to have blooded in due form both Captain Thomson and Major Whyte Melville — an afternoon's work well worth remembering. Stravithie is also a very strong and noted gorse of twenty-six acres, and the finest woodlands are Melville, Falkland, Airdrie, the Earl of Leven's woods, Dhu Craigs, &c.

Captain Thomson generally kept about forty-two couple of working hounds, and ran them mixed. Charleton, New Inn, and Torriburn were all stations, and much they were needed; but Cupar was desolate. The old kennels form part of the coal depôt at the railway, the flags are in the Charleton kennel, and old Shepherd, the feeder, is still in commission at New Inn. The hounds were out in the meadow behind, and a waive from the Captain would have sent them off in a crack to make inquiries at Nottingham. Old age was creeping on the grey-eyed Rallywood; Ornament, half-brother to the late Tom Sebright's beloved Ottoman, looked as if he had been "boxing a bit;" and Ravisher of the delicate nose had got it all scratched in puzzling out his fox the evening before, among the ivied ruins of Balmerino. Ravisher and Ransom by Wemyss's Ringwood were the only Blossoms left out of the seventeen, and have all the fine, low-scented properties of their grandsire, Drake's Duster. There, too, were Bajazet, Bondsman, and Bonny Lass; but Baronet, the best of the litter, was found dead near the railway bridge at Perth, and there had been as much lamentation over him as there was for Syren. Bonny Lass, one of Fred's "premium lasses" at Redcar, was of course called up to verify what so much astonished Captain Williams and his tape-line, viz., that she is $29\frac{1}{2}$ inches round the heart and 5 inches below the knee.

Dairymaid was another of the Redcar three couple, and as good as an otter hound among the rocks; and Tempest, Charmer, Symmetry (a niece of Syren's), and Tragedy made up the prize lot, which had a long seaside ramble on the Redcar morning, and came to the post as clean as smelts. Conqueror is a grenadier to look at, but always there when there is a pace; so is Ringwood, one of the big Bramham blue and whites, hard in temper, but capital in work, and the sire of a capital litter from Captive. Matron is a regular chub, and Melody all life, with too short ears. "*She's a merry beast—all the Marlboroughs were,*" said Fred, enunciating this great truth with a deeply solemn face. Then the Master chimes in: "*As for Trooper, if Fife were not almost an island, he could'nt have been kept in it, though he goes out every day.*" Ranter, has been out thirteen times running; but the sort are all hard tempers, and difficult to break, and "the very flesh has to be knocked off their bones to keep them quiet."

And so we run through them. There is the red Wildfire, so often in front that she has been mistaken for a fox; Rhoderick, the low and thick line-hunter, who runs in the middle of the pack, and of course kills the foxes; Sportsman and Songstress, the last of the Syrens, good in themselves, and loved for her sake; Struggler and Striver, lathy, but full of drive; little Damsel, light over the chine, but never idle; Favourite from Fashion, whom Captain Williams declared to be the best; Barmaid, simply "a trimmer;" the half-faced Reginald, "about our leader, who'll turn at a mile for his master's whistle;" Standard, "my best friend," though he has one eye, and his toes all but broken by a Rufford trap; Rhapsody, small but very pretty; Ransom, who can carry it further along a road than any of them; and Mystery, who was buried in a sand-bank for nearly half-an-hour, and was speechless when she was rescued, and yet scratched up to her fox again, and held him till they were dug out.

For slow work Standard and Bajazet were the best, and Struggler and Reginald quite the leaders of "the guides." Standard was entered by Oxtoby, and narrowly escaped being hung for roe deer; and Winsome was steady in her devotion to Lord Rosslyn. She was walked at Dysart, and would never leave his lordship, except for a fox, and then finish 300 yards up a drain at times.

The last day of the Captain's reign was at Kilgraston. They found at Glencairn, and ran to Invermay, but they could hold the line no longer in such dry weather. However, they drew Glencairn again, and gave it up after a couple of hours. Bonny Lass and Tempest died the same evening. Both were full of puppies, and there was an ulcerated ring round the neck of every one of them, which puzzles the profession to this hour.

CHAPTER XVI.
ALLOA TO SKYE.

"What bliss and life can autumn yield,
 If gloom, and showers, and storms prevail;
 And Ceres flies the naked field,
 And fruits, and flowers, and Phœbus fail?"

The Wood of Caledon Bulls—The Duke of Montrose's Herd—Sail to Skye—Products of Portree—Ride to Duntulm—Cattle and Sheep in Skye—Skye Terriers—A Pig Hunt—Symptoms of Falkirk—The Poltalloch Herd.

Autumn drew on, and we left for a time the great Wood of Caledon. It ran originally "fra Strireling (Stirling) throw Menteith and Stratherne to Atholl and Lochquabir," and was inhabited by gritrit bullis with crisp and curland mane, with sich hatrent aganis ye societe and company of men."

Only a few of these "bullis" are left near Hamilton. They are almost universally white, with black ears, muzzles, and feet (points in which the Chillingham are red), and generally horned. If they come polled, it is always considered a mark of bad blood. They are of good medium size, and compact in form. One of the patriarchs of the herd, who was shot about five years since, measured two feet from the frontlet to the tip of the nose, or the same as the span between his horns, which were ten and a-half inches in height. The length from his neck-vein to the root of his tail was five-feet-eight; so that with these data and a bone a Professor of Geology should have no difficulty in building one to order. Some are generally killed every year for the poor; but they have been occasionally used, and liked, in Hamilton Palace. They are always shot, and fetched away after the commotion occasioned by the fall has subsided; but stalking them is no easy task. The bulls go in front, with the leader in the centre, and the calves between them and the cows; and if at all pressed, they come thundering on like a charge of Life Guards. A young calf has been sometimes found by itself, and carried off to the farm to be fed; but it is a perilous task, and the calf begins butting at two days' old, and seldom grows milder with handling.

As for the Wood of Caledon, the Romans are reputed to have cut it down, to drive out the Celts; and oak-trees have been found, with canoes and the remains of a whale, in the Vale of Monteith, as well as Flanders Moss. Once upon a time, there was six to fifteen feet depth of bog earth, but it has been floated away, and the fine clay beneath forms the surface of a carse which chiefly grows wheat and beans. About Strathblane the scene changes to the old grass of the dairy districts, which keep Ayrshires for the supply of Glasgow, and send in their milk tubs morning and evening by rail to the Clyde side at Dumbarton. Buchanan Castle, the seat of the Duke of Montrose, is about two miles from Drymen Station on this line. His Grace once had nothing but Ayrshires, and it was not until 1857 that he ever purchased a shorthorn. The sales at Neasdon and Bushey tempted him with two roans, Primula and Doraliso, but he had not much satisfaction in his new pursuit till he bought New Year's Morn by Baltic as a 60-guinea calf at Mr. Cator's sale. The right chord was struck at last, and her very first calf, May Morn by Victor Emanuel (15460), a bull purchased from Lord Feversham, was head in the Royal two-year-old class at Battersea. Five or six years had swelled the prize-list to twenty-eight, and the herd to thirty-one; and the three Morns—New Year's, Rosy, and May—along with Baroness Killerby and Flower Girl, were the only ones which were not brought under Mr. Wetherell's hammer. Old Flavia by Baron Warlaby contributed her 62 guineas towards the 35-guinea average, and her granddaughter Lucilla by Baron Killerby followed Mr. Young's nod to Keir. The herd had a narrow escape from pleura, which began with a Bretonne in a byre adjoining the Ayrshires. Every one of them fell ill; but the shorthorns, which were in another house, with a dunghill between, and were then removed for safety to the stables at the Castle, did not suffer in the least, and added another to the mass of proofs in favour of infection *versus* epidemic.

The steading lies snugly a few yards to the right, as you drive up the avenue towards Bu-

chanan Castle; the Perthshire Hills and Ben
Lomond rise in the distance; and Loch Lomond
and Loch Katrine are not far away. But we care
for none of these things, and only long for an "interior," such as John Taylor has to show, as he
throws back the door of the boxes in turn, and unveils New Year's Morn. She seemed as jealous of
her calf as a cow well may be, when she has bred a
Royal winner. Hers and Flavia's are the tribes of
the herd; and Rosy Morn and her bull calf, May
Morn, and Morning Star, range themselves under
the one, and Flower Girl and Baroness Killerby
under the other; while Rosedale and La Valliere are
the outsiders. Fashion by Baron Killerby from Flavia
had just been sold to Mr. John M'Kessack; and as
only two bull-calves arrived last year, there were, including calves, only fourteen in all.

Rosy Morn by Victor Royal (21028) stood next to her
dam, and had a good Fashion bull-calf; and then came
Flower Girl, the first-prize heifer at Warwick, with
her second calf at her foot, and not very much milk
for it. She has held herself together very fairly, but
her look is rather spoilt by the small horns, which
she inherits from her grandam. Baroness Killerby
has the same peculiarity, and, like the other three, she
is full of rich hair, but her roan Fashion calf is dead.
The same ill luck had befallen Rosedale, whose bull-
calf died suddenly at the end of twelve weeks. She
herself did Taylor a world of credit for the careful
way in which he had reduced her, and she looked all

that a cow should in point of condition, and, better still, safe in calf again to Fashion. May Morn has made no way since Battersea; even a Highland bull had been in vain, and she was under orders for the shambles at last. Once she was not in season for six weeks, but generally on every 11th or 21st day. It turned out that there was an internal malformation, which rendered breeding impossible, and so the overfeeding hypothesis falls to the ground. White prize heifers seem quite unlucky in Scotland, as May Morn, Clarionet, Venus de Medicis, and Lady Windsor were all under this ban together. La Valliere had just had a heifer calf in her old age, and one of the sweetest heads that was ever put on calf belonged to her young Scottish Chief by Ravenspur. It haunted us as we retraced our steps to the Drymen station, and long after we had taken the boat at Dumbarton for Greenock, and were fairly booked for Skye by the "Deep Sea Sailings."

"The Drum" was up at Greenock, but it was no prophet for the West Coast; and the captain bade us be of good cheer, and take our rest round the Mull of Kintyre. There was no tossing save in dreams; and the white crescents of Oban, dear to tourists and college reading parties, greeted us when the breakfast-bell rung next morning.

Cheddar cheese was the burden of the song, when we asked about Coll and Kintyre. There was no cheese-making in the latter previous to 1831. It then became so certain that the soil did not suit West Highlanders, that Mr. John Lorn Steuart of Coll, chamberlain to the Duke of Argyll, began to keep Ayrshire cattle; and ever since then dairy husbandry has been universally adopted. The farmers' daughters came to be instructed in the making of Dunlop cheese, but still Mr. Steuart was not fully satisfied with the result; and feeling con- convinced that the goodness of the cheese was more dependent upon the making than the pasture, he sent two of his servants to England to learn the Cheddar process, which was then hardly, if at all, known in Scotland. Last year he purchased the celebrated prize-bull " Sir Colin Campbell" for his Grace the Duke of Argyll, who wished to improve the breed of Ayrshires among his tenantry, and about sixty calves for the present season is no bad beginning. Mr. Steuart has owned the Island of Coll since '56; and in- troduced the dairy system on the home farm with a herd of eighty cows. The cheese is of finer quality than the Kintyre, as the sandy downs are covered with the richest clover; and in 1864 it brought no less than £66 per ton, the highest price that has ever been paid, to our knowledge, for Scottish Ched- dar in the London market.

We were soon on our way once more, past that time-eaten keep, tapestried with ivy, and crops of oats waving on such haggard, sea-bound spots, that, even if we do spy a house in the distance, we lose the idea of the Lord of the Isle in " the man that couldn't get warm." Sometimes we are steaming up a bay to an Argyleshire pier; or perhaps we merely stop in the offing, and a boat is sent off to us. An old Corunna man in one of them is quite a god-send in the dulness. He is at once invited down to the principal cabin to show his medal, and gets a second glass on the strength of his own, as well as Sir John Moore's memory. Occasionally we skirt the open Atlantic, with the usual result; but when the evening is far spent, we are land-locked close by Isle Ornsay, among the very intricate navigation, and wait patiently for the moon to rise. Its beams light up the shadowy oriel of the U. P. church, as we scramble out of the boat at Portree; but it takes many a thundering knock to rouse the inn. The cattle show has come off the day before; and the convener, vice-convener, and all the rest of them, are snoring in their dreams.

Portree is not lively by daylight. The principal shop seems to unite castor-oil, senna, and Harvey's sauce, with "two practical discourses," and photographs of the man in shoe-buckles who made them. There are cakes elsewhere, of a texture which goes far to prove that Young Portree must have the stomach of a cock. The only visible remnants of the cattle show are a few loose hurdles in a meadow; and we find the pick of the Duntulm Highlanders and Cheviot shearlings, all with first-prize cards on their heads, browsing near Kingsburgh Bridge in the middle of their twenty-mile walk. The milestones point to no towns, but only to inns; and there is a sort of benighted feeling as you look at them, with nothing but mainland all round you. *" Snipes appear,"* is the entry for the day in " Cuthbert Johnson's Farmers' Almanac "; and certainly there is a fine opening for them in Skye, as three weeks of continuous rain had left suction in abundance.

Where the rooks go to at night is our great puzzle, as there are only a few brown ashes to be seen, but Skeabost Woods is their Aldershot. Its direction is indicated by our friend somewhere to the left beyond Loch Snizort, on whose blue waters whole navies might ride. A pinnace is at anchor behind the breakwater in Fraser's bay—the snuggest spot we have seen yet, with Ericstane tups in the pastures, and a library we might winter on. Ben Edera's snowy head warns the cottars that they must gather in their oats; but while they work at their scanty sheaves in one part of the field, the cow and the keeries are alike busy at the other. The cow is wonderfully ubiquitous. If the man is plaiting heather ropes, to keep the thatch on, she is ruminating at his side; if you go into a shop, it is anything but certain that she may not look over the counter. Some cottars have as many as 16 to 20 keeries, generally black and brown-faced. They kill and salt the wedders and old ewes, and the lambs come in from the hill and walk about with the family. Strangers are generally told, in confidence, of a wedder which was salted in the seventeenth year of its age. Not a goose or turkey is to be seen, but pigs occasionally; and, in fact, oatcake, cuddy, and lyth, which they fish for with a line and swivel, and a skinned black eel, are what the cottars mainly trust to. There is that low, sighing wind, which betokens abundance of rain; and the hay-fields are soon the scene of one great Scurry Stakes among the women who carry half a hay-cock at least on their backs to the rick.

Quirang is in sight at last, with its chain of natural ramparts, the glory of which would make the sternest of Woolwich martinets forget himself, and play at leap-frog with unbuttoned jacket and cadets upon the green knolls below. Over the water is the shore of Ross-shire once more, with its eternal cottars. They fish, and they live as they can, and multiply like the eight-year-old black mare which trots away over the heather with five blacks, all her own, and none of them twins, at her heels.

Duntulm Castle looks bleak and bare, as we visit it next morning. Its days of revelry have long gone by, but it has been orally handed down that there was dancing in it about two hundred years ago. Now the witches and warlocks have all the reels to themselves. Black-faced wedders browse in the old garden, and an empty cask, with sad suggestiveness, was tossing about in the dungeon, where nothing passed the lips but salt boiled beef and hopeless cries for water.

Skye is divided into seven parishes, one of which, Kilmuir, is on clay, and the rest on good loam. The finest grazing is at M'Leod of M'Leod's, and also in the parish of Bracadale. There is very little high farming to speak of. Regents and Irish Rocks are the staple of a grand potato crop; turnips are few, and mangels unknown; and all the wheat flour is imported. Except at a stall in Portree, there is very little public beef and mutton. Butter and cheese are not exported, and all groceries come from Glasgow. The horses have been crossed in-and-in with half Clydesdales and ponies, and now they seem to have settled into a large fourteen-hand pony, which would be none the worse for another stain of blood. North and South Uist are the pony islands; and both of them used to have races on the sands, by way of celebrating their harvest-home at Michaelmas (old style), with dancing and Michaelmas bannocks to follow. These "struans," as they call the cake of barley-meal in the shape of a heart, are toasted before the fire, and dressed with treacle, eggs, and carraway-seeds, and then eaten by lovers and guests in general.

We did not go on to Lews, as, with the exception of its smaller breed of cattle — black with brown backs and ears, killing from 300 to 350lbs. neat—the flock and herd history of one of these islands is pretty much that of another. Lewis or Lews is generally flat and mossy, with the exception of the south end of the Island. There are splendid hills and glens abounding with deer and grouse and other varieties of game. The ancient

Forest, whose "park" is let to Mr. Sellar for sheep pasture and shooting, was the regular shooting ground of the proprietor of the island, and for thirty years Archibald and his brother Alexander Stewart were its tacksmen, and reared first-rate black cattle and black-faces on it. The rod-fishing throughout Lews is far superior to anything of the kind in the Hebrides or Skye. Its sole proprietor, Sir James Matheson, M.P., has improved the island to a great extent by roads and bridges, and established several schools. He has also done much to improve the native cattle by the introduction of West Highland bulls, and brought over thoroughbred sires and Arabs as well. Some of the smaller tenants have improved black-faces, but the Cheviots are gradually putting out the small Island breed of sheep which are brought to the Stornoway market, and the best of which only kill when they come to three-year-old wedder estate, from 7lbs. to 10lbs. a quarter, and clip from 2lbs. to 3lbs.

Of "The Seven Hunters," that mysterious group to the North, we do not presume to speak. Macdonald of Balranald has a large herd of big West Highlanders in North Uist; but the cattle in the Uists are not thought quite equal to those in Barra. In Skye, Macdonald of Waternish, Mackinnon of Corrie, Nichol Martin of Glendale, Macrae of Knock, Captain Cameron of Talisker, and Stewart of Duntulm, have all large Highland herds. Stewart of Ellanriach got a gold medal for the best cow at Paris, but he has not shown of late years; and Stewart of Duntulm never shows on the main land, but has fairly won the championship with his "Targill" stock. One of the finest family bulls was out of "Lucky," which was second at Inverness in '46;* and after winning the local Highland Society's prize as a three-year-old the next year, carried it for old bulls in 1848 at Slegichan, against a celebrated yellow bull of Sir Robert Menzies' breeding.

All the bulls and cows have Gaelic names here —signifying "White Tuft in the Tail," "Piley," "Beauty," "Nice," and so on. The calves are weaned at the end of five months, and are kept in a loose box all winter, but the cows are never in-doors, and their summer consorts are removed to "parks." Small tenants who have half-a-dozen cows and "followers," join to hire or buy a bull; and sell their stirks at a year old for £3 or £4 to dealers who come before September Falkirk, and take them across the ferry or by the steamer. The farmers who have enough keep, hold on for another year, and get their £6 to £11. There is hardly any veal, except the shepherd's cross-bred quey has a bull calf, and then it is killed off at six weeks. Once there was plenty of kid flesh and sweet goat cheese; but the white, black, and grey lichen

---

* We should have said at page 83 that Inverness "has already taken its turn three times in the Highland Society's circuit." This is of course exclusive of the '65 meeting. At page 58 the terms of the silver medal contest are not correctly stated. *Tenant-farmers* can only compete with a tup which has won a first prize at a previous competition.

croppers have nearly all disappeared, save at Mackinnon's of Corrie.

The sheep farms are generally held on leases of from fifteen to nineteen years. Half the island is under sheep, and its sad lack of straths is very slightly compensated by a sweet bite along the streams. There is no luxuriant heather, and the climate is too wet for lime-dressing. The sheep drains are eighteen inches in width and depth. Blackfaces have gradually retreated before the Cheviots; but Macdonald of Skirnish is strong in the old faith. He gets his tups from Lanarkshire, and his " crock ewes" are sold at five years old at Falkirk or Inverness character market. Scott of Drynoch is one of the largest Cheviot men, and he generally keeps and sells his wedders at Falkirk. Nearly all the farmers sell wedder lambs, and get them away before Aug. 12. Dealers come round on the Monday before the September fair at Portree, and buy by the clad score, but there is no pitched sheep market in the islands. There is a good deal of sturdy, but flockmasters generally prefer taking it short, and " take the head off instead of probing for the blot." Braxy is terribly severe upon the hoggs; and if they were not sent away into Ross-shire, Inverness-shire, and the Black Isle to turnips, no farm could keep itself in stock. Even the Cheviot breeders have a taste for blackface, and buy a few wedders from the smaller tenants to put on to islands and parks, and kill them one by one as the giant did the children. The ewes are generally milked for eight or ten days, and the milk mixed with some from the cow; so that the gude wife can hardly put the same query to you which we have heard in Roxburghshire, after the second or third course—" *Yow or cow?*" It is a great point to make the cheese as soon as possible after the milking, and one hundred ewes go to an eight-pound cheese.

There are no hounds, except the few which are used by the todhunter, who is supported by an assessment on the tacksmen. Mr. Mackinnon of Kylaackan is fond of good Skye terriers; and Captain Macdonald of Waternish keeps a regular pack for otters, which abound all over the coast, and make large levies on the Snizort salmon. The mode of hunting would horrify Mr. Waldron Hill or Dr. Grant, those great Scottish representatives of the sport, as they put them out of the cairns with terriers, and then shoot them.

Many of the so-called Skye terriers are not "pure Skyes," but a cross between that breed and a French poodle, of which two or three specimens swam on shore when a ship was wrecked near the coast. Through them the original, short, wiry-haired dog was changed, in a great measure, into one with a long silky coat, and hence the white, long-haired terrier which was at one period so common in the island. By this cross, the properties of the dog have been quite changed, as he is not a fighter like the pure Skye, and only used for tracking vermin. A friend of ours in Aberdeenshire bred and tried sixteen of the sort, and only three of them would face a polecat.

Still, they are superior to the long-haired curs which are reared by the poor people round Portree and Broadford, to supply the wants of tourists and others, who expect they are buying pure Skyes, and only get pretty toy-dogs. The true Skye is a long-backed, short-legged dog, with wiry hair, ears generally drooping at the point, and weighing about 16lbs. The orthodox colour is dark grey, and the breed is rarely to be got except from some of the gentlemen of the county. They are dogs of extraordinary pluck, and will "go straight at" an otter, wild-cat, or fox, after tracking them into their deepest dens among the rocks and cairns. They will bear an immense amount of punishment, and not unfrequently never live to come back when they have tackled something quite above their weight.

Crows, ravens, and wood-pigeons live in the rock clefts, and have, on the whole, rather a harassed time of it; and so have the golden-yellow fishing eagles. The deer keep to the Coolin hills and forests; a few pheasants flourish at Armadale and Dunvegan; grouse disease is unknown, and hares never turn white. The most extraordinary, and in fact the only, hunt we saw was on a Sunday. The sheep dogs were all lying at their masters' feet, while the clergyman delivered a sermon in Erse, of which we only understood the " Armen." Bacon is rarely eaten in these regions, and the dogs, from some cause or other, view the roving pigs with the deepest disgust. One of them sauntered into the little school-room in the course of the service, and in an instant the whole pack were on foot—a gallant black, which had sat with its paws on our knees or round our neck for a considerable portion of the service, in affectionate remembrance of two butter pats at Portree—getting away well at the hams. We heard them carrying a tremendous head over the peat bogs and through the oats. Two or three of the minister's sentences were completely drowned in the cry; but he held calmly on, neither looking to the right or left, with the air of a man thoroughly used to such finds and tremendous bursts in the open. As for the little old woman in the high cap, who sat with her face to the wall, she never moved a muscle.

We walked back with the minister, who strode along in his hoddan grey suit, with a pastoral crook to aid him. A strange time we had of it, among wind, showers, and rainbows—now dodging behind a peat stack, now under a wall, and hearing for our comfort that "*the wet comes here in cycles.*" He had been brought up in a stern school, when, as he phrased it, "Latin was flying about at many a table," and when, with a gillie to carry his luggage, he had to walk his two hundred miles as a student, to pluck the tree of knowledge at St. Andrews. He told us a very sad tale of the destitution of his poor parishioners in the winter; and how many of the children can never come either to school or church, but crouch round the embers with hardly a rag to cover their nakedness. A Skye minister's life must be a weary one; and an old tourist thus quaintly expressed his sense of it: "I'm so sorry for them, that, whenever I see a *Times* newspaper lying about, I always direct it off to them at once." The church at Kilmuir looks like a barn on a bleak headland, and is at least two miles away from the graveyard where Flora Macdonald, who left five soldier sons, sleeps the sleep of five-and-seventy years.* The marble slab has fallen out of the stone, and has been carried away, piece by piece, for relics; but nettles and ragweed thrive right bravely. We turn aside to look at the Balaclava-like plains behind, to read the virtues of a minister's wife, and of a knight at rest with very short legs; and then pass on our way back to Portree. Again it is all wild muirland, among which stands the new Free Kirk and its manse,

---

* The following interesting note on the subject of this grave occurs in Mr. Robert Carruthers's illustrated edition of Boswell's "Journal to the Hebrides":—

"Flora died on the 4th of March, 1790, aged 68, and was interred in the churchyard of Kilmuir, in a spot set apart for the graves of the Kingsburgh family. The funeral was attended by about three thousand persons, all of whom were served with refreshments in the old Highland fashion. Kingsburgh died on the 20th September, 1795. Flora had seven children, five sons and two daughters; the sons all became officers in the army, and the daughters officers' wives. The last surviving member of this family, Mrs. Major MacLeod, died at Stein, in Skye, in 1834, leaving a daughter, Miss Mary MacLeod, who resides in the same place. One of the sons, the late Colonel Macdonald, of Exeter, sent a marble slab, suitably inscribed, to be placed near his mother's remains, to point out the spot; but it was broken ere it reached Skye, and the whole has since been carried off piecemeal by tourists. Thus the grave of Flora Macdonald remains undistinguished within the rude inclosure that holds the dust of so many of the brave Kingsburgh family."

looking as if they had been built by a wandering band of free and accepted masons, or let down by invisible hands.

The sea journey back by The Clansman was rather more lively; and again we passed Scalpa and Raasay, in whose Sound Dr. Johnson lost his spurs over the side of the boat.

> "O'er them mouldering,
> The lonely sea bird crosses
> With one waft of its wing;"

and we thought, if we could only dive and find them encrusted with *molluscæ*, what would'nt they fetch at Christie's? Falkirk October is at hand, and our principal mission seems to be sailing up and down bays, and shipping sheep and wool-packs. The forecastle is quite a parliament of Cheviot and and keeries of all colours, every twentieth with horns as a standard-bearer. Cows, half shorthorn and half Ayrshire, are pushed in from the piers, and a blue-grey pony, which would have been a perfect palette puzzle for Herring, is boxed up among hampers containing a beagle and two black-and-tan setter puppies for a Hampshire vicar. The two visitors whom we left at Eigg on Friday seem to glory in their release; but barren as the island looks, hazel bushes at the prow of the boat that brought them and the farmer and his eight-score ewes, told of bosky dingles somewhere. Cheviots and cattle are its products, and its cottars seek the lowland harvest; whereas Rum, which is a wilder island, goes for deer and blackfaces entirely. "More wool, a cross-bred bull (owner tries to make him out a pure Shorthorn, but his man won't have it), more lobsters, and brides and bridegrooms" is our note as the day wears on. The greatest pleasure is to lounge on deck with Hugh Miller's "Cruise of the Betsy," and read through his eyes, rather than our own, the outline of these "fractured Caledonian Isles." There is the "one low hill" of Muck; the "pyramidal mountains of Rum, grey in fog and sad in rain, in whose wild hollow the withered female is seen before death in the twilight, and washes a shroud in the stream"; and Eigg, with its "colossal ridge rising between us and the sky, as a piece of the Babylonish wall or the great Wall of China."

The coffee-rooms at Oban look warm and cheery, and, with the leader of the Parliamentary Bar among the passengers, we are off again in the morning towards the Crinan Canal, where the four-horse boat stands ready, with its postillions in scarlet and velvet caps. No mode of locomotion is like it, and we might well be loath to get out at the first lock, and take a long walk over the moss; but Kilmartin Glen and Poltalloch "atone for all."

West Highlanders still hold a large portion of the peat moss, but year by year their old domain decreases, and purple-tops are in the ascendant. Twenty years ago, Mr. Malcolm adopted wedge drains, but now only pipes and collars are used, about eighteen feet apart. On moss land the drains are 5 ft. deep, and on sound 3 ft. 6 in. to 4 ft. The former is ploughed five inches deep after draining, and then sown with oats. After the next ploughing, lime is harrowed in before ridging, and it is manured with 1 to 2 cwt. of superphosphate per acre, as well as farmyard dung. Under such treatment, as the bailiff William Stewart expresses it, the peat may well "melt like snuff."

"The Black Cattle," as West Highlanders are always termed in Argyllshire, have a title of fully seventy years at Duntroon. The first bull of note was purchased about five years later, along with several cows, from a herd in the island of Shuna; and those of Major M'Donald of Corrie Broadford, Isle of Skye, M'Donald of Monachyle, Stewart of Duntulm, M'Claren of Camuseroch, and the late Marquis of Breadalbane (from whom Crinan was bought) have all furnished "kings in their turn." Forty years since, some choice cows and heifers came from George Sixth Duke of Argyll, and more recently from the Breadalbane sale, and Mr. John Campbell's of Lochead. There are about fifty cows in all; and a glimpse from the steamer deck of Duntroon, the brindled bull, as he wondered along the sh ore near Duntroon castle, with the dun cow' (which also won at Battersea) and a calf at her side, recalled those "bull, cow, and offspring" groups which Herefordshire loves so dearly. Mr. Malcolm enters pretty freely into showing, and the Highland Society has no steadier adherent each August. Some very choice ones are in his herd, but they can only be described by painting; and the "Return to their Native Heath of the Winners at Battersea," by Mr. Gourlay Steell, who renews his strength in Argyllshire as each summer comes round, best tells the story of their shaggy coat and tameless eye. All the cattle except the five-year-old heifers are wintered in the house. Sexton boars have also worked their way down here, and lie stretched on fern in their iron and Caithness flag tabernacles. Berkshires are kept, but they are not liked nearly so much as the Essex, and, in fact, the taste for pigs finds no "lateral extension" in the county. Ploughmen do not care to keep them, and would think very little of half-a-dozen by the side of a "bit of an Ayrshire."

The black-faces have flourished for only six summers at Poltalloch, and the flock numbers, on an average, about nine hundred to a thousand. They succeeded Leicesters, which, like the Downs, found the climate too wet, and black-faces have proved the masters of the situation. The "curly horns" were principally selected by Mr. Martin, the factor, from most of the prize-winning flocks in Scotland. Then the third-prize shearling at Kelso lent his aid; and the shearling with which Mr. Malcolm himself bore off the same honours at Battersea was sold at the annual roup for £20 11s. Nearly all the farmers round are purchasers, and the highest average as yet has been £5 16s. 8d. for twenty-two. The horses are generally of the larger Highland sort, and one old mare

www.ingramcontent.com/pod-product-compliance
Lightning Source LLC
Chambersburg PA
CBHW032017220426
43664CB00006B/276